MASONRY AND CONCRETE

MASONRY AND CONCRETE

Byron W. Maguire

RESTON PUBLISHING COMPANY, INC.
A Prentice-Hall Company
Reston, Virginia

Library of Congress Cataloging in Publication Data

Maguire, Byron W.
Masonry and concrete.

Bibliography: p.
Includes index.
1. Masonry. 2. Concrete construction. 3. Blueprints. I. Title.
TH1199.M23 693 77–28522
ISBN 0–87909–521–0

Reston Publishing Company, Inc.
A Prentice-Hall Company
Reston, Virginia 22090

10 9 8 7 6 5 4 3 2 1

PRINTED IN THE UNITED STATES OF AMERICA

Contents

Preface

Each book written on one of the building trades usually deals with one, two, or three particular aspects of a trade. This book focuses on the importance of the blueprint to a mason. The blueprint relates to the experienced journeyman, a considerable variety of factors, data instructions, and principles of masonry. From the combined information the mason is able to select appropriate materials for the job, construct logical sequences of work progression, and know where and how to install the materials.

The book is divided into four sections; Section I is devoted to an understanding of concrete and its many uses. Section II develops an understanding of the principles of masonry using bricks, blocks, and stone. Section III deals with concrete and related materials used to create artistic forms. Section IV covers various repair and maintenance tasks.

Since so many different requirements exist for making, for example, footings, or walls, or even patios, each section is broken into chapters. This organization allows the reader to focus attention on just one segment of masonry. In Chapter 2, for example, footings are studied. Blueprint data are provided, identified, examined, and explained.

Where blueprint data are lacking or are not usually included, specifications are supplied by the architect. These, too, are included in this chapter and others. Other relationships are also provided, including; sequences of tasks needed to complete a job, fundamentals of tasks used by masons and carpenters, and an inspection checklist that identifies quality workmanship.

Not all masonry is crude, rough, and utilitarian. Much fine artistic work is done by experienced masons, and this must be understood and learned by the novice or apprentice. The simple task of striking joints adds artistic quality to a wall. The not-so-simple task of laying bricks or blocks to create precise patterns is also an artistic application of a common material. Then, too, designs and textures are made into sidewalks, driveways, patios, and the like. Throughout the book, details are shown and explained so that the reader may understand how to make the results more beautiful.

Art using masonry materials does not always need to be developed from blueprints and specifications prepared by professionals. Chapters 14 and 15 develop ideas and plans that provide readers with the opportunity to develop creations of their own design and making. Several principles of idea development are given, along with a flowchart of production and a method that is useful in estimating materials and time. Chapter 15 develops several artistic projects. These are detailed rather elaborately so that readers can understand the author's thinking as they were being developed. The projects can be used as a starting point for the readers' own creations.

* * *

Thanks are extended to the following contributors. Their data contributed to the depth of subject development and provided greater understanding of both the principles of masonry and the way the tasks are carried out.

American Concrete Institute, Detroit, Michigan.
American Plywood Association, Tacoma, Washington.
Brick Institute of America, McLean, Virginia.
Goldblatt Tool Co., Asage, Kansas.
Marshalltown Trowel Co., Marshalltown, Iowa.
Masonry Institute of America, Los Angeles, California.
Portland Cement Association, Skokie, Illinois.
Stanley Tools, New Britain, Connecticut.

MASONRY AND CONCRETE

Section I

Concrete and Its Many Uses

In each of the eight chapters of Section I we examine a different application of concrete. Before a knowledge of the use of concrete can be gained, it is important that concrete and its properties be studied. Chapter 1 examines the properties of concrete. These include the hydraulics of concrete, various mixtures, and curing, to name a few. Following are chapters devoted to concrete uses in all sections of a building and surrounding areas. Starting with footings, the study focuses on translating blueprint data to reality. Following chapters cover concrete used in walls, floors, stairs, sidewalks, and patios.

A great deal of emphasis is placed upon the use of blueprints and specifications. The rationale behind this view is that each part of a plan or each specification communicates to masons. They do not see lines and numbers but see a series of activities associated with each drawing detail. From what seems to be a very limited amount of data, they are able to construct a sound structure.

How do they do this? Each mark, line, or notation on a plan has a special meaning. Figure A illustrates several drawing symbols found on a blueprint.

1. The main lines outlining the structure are bold, unbroken lines. These are found in such areas as foundation wall, elevation, and sectional drawings.
2. Dimensional and extension lines are lightweight unbroken lines that show the distance between two points. The dimensional lines usually have arrows or dots at either end and have length or height notations either above the line or midway within the line. Extension lines are perpendicular to dimension lines and extend outward from main lines.
3. Lightweight broken lines are used to indicate that part of a wall or other detail has been omitted.

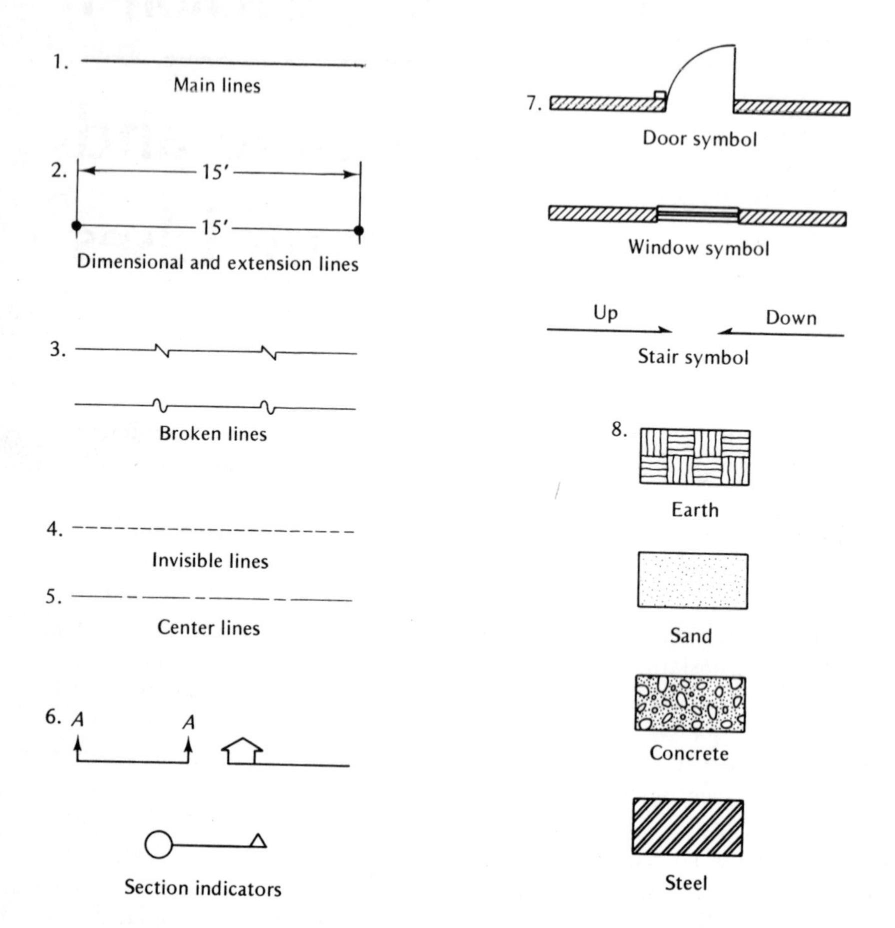

4. Invisible lines, made from a series of short dashes, are used to show portions of the building that are below the surface shown. Footings are an example of such lines.
5. Center lines are constructed of lightweight alternating long and short dashes and are used to locate centers.
6. Section indicators are lightweight lines that have one of a variety of symbols at their ends. The symbols at the ends refer to sectional drawings that contain the full or partial cross section of the wall.
7. Doors and windows and stairs are also indicated by symbols.
8. Earth, sand, concrete, and steel are shown where applicable, but usually in sectional or detail drawings.

From time to time, readers begin a study but do not exactly know their main goal. For this reason, the objective of each of the following chapters is identified in the opening paragraph. By careful reading of the objective a purpose for study is centralized, and concentration can be fully directed. Quite frequently there are two objectives per chapter. The first points to a need to understand the blueprint data. The second directs attention to the practical tasks necessary for building the object.

Each chapter closes with a quality-control inspection list. These simple lists define the various phases of the work where the best workmanship must be used. Each item in the checklist deals with a small segment of the total task. The sequence of items listed follows the logical progression of the work. Therefore, the beginners and apprentice masons may find the checklists very useful as preinspection lists to assure that they understand the points that must have the greatest accuracy.

Chapter I

Concrete - What Is It?

Admixtures all materials, other than portland cement, water, and aggregates, that are added to concrete, mortar, or grout immediately before or during mixing.

Aggregate bulk materials, 'such as sand, gravel, crushed stone, slag, pumice, scoria, and vermiculite, used in making concrete.

Air entraining an admixture added to concrete to improve its durability.

Coloring agents colored aggregates or mineral oxides ground finer than cement. White portland cement is also used to effect color changes.

Curing methods employed in allowing the hydration process to complete its cycle.

Hydration the cementing properties due to the chemical reactions between cement and water.

Lightweight concrete a type of structural concrete that has an air-dried unit weight of less than 115 pounds per cubic foot.

Mixers vehicles or containers used to blend or mix the ingredients of concrete.

Moisture content the amount of water contained within the aggregate used in concrete.

Paste a mixture of cement and water.

Plastic consistency sluggish flow without segregation.

Portland cement a manufactured product of mineral substances that have hydraulic properties.

Thermal conductivity the property of concrete that allows for conduction of heat through its structure.

Thermal insulation the insulation of lightweight concrete, which resists the conduction and penetration of heat.

Trial mixtures specially measured quantities of cement, aggregate, and water, or mortar mix that provide data on quality, workability, and economy.

Water/cement ratios a decimal calculation based on the water/cement ratio, pound per pound.

White cement portland cement, made from minerals excluding iron and manganese oxide.

Workability the ease or difficulty of placing and consolidating concrete.

OBJECTIVE—INTRODUCTION

The terms listed above should be studied carefully, for they describe the substance and the nature of concrete. However, it is equally important that a study have an objective, for it focuses attention on the primary subject. Since a beginning on the study of masonry is being made, the first objective is that *each reader be able to select the proper materials and combine them in proper proportion to make concrete.*

A simple chart showing the proper proportions of fine aggregate (sand), coarse aggregate (gravel), cement, and water would suffice for mixing instructions. Such mixes as 1:2:3 and 1:2:4 are common, their numbers representing (in order) cement, sand, and gravel.

These figures do not, however, give the complete picture. The serious student needs additional data, such as the physical properties of the ingredients making concrete, concrete's chemical properties, the types of mixes suited to various needs and jobs, and mixing methods.

PHYSICAL PROPERTIES

Aggregates are classified into the subcategories: *fine* and *coarse.* Fine aggregates consist of natural or manufactured sand with particle sizes smaller than ¼ in. They play a very significant role in the formation of concrete, in that they assist in filling the spacing between the coarse aggregate and aid in bonding. Since sand is almost always rounded or cubical, it aids in the workability of the mixture. Observe in Figure 1–1 that the fine aggregate, sand, accounts for 24 to 30 percent of the volume of concrete.

Fine aggregates made from shale, clay, slate, slag, and vermiculite are manufactured in expanded form, where they are then crushed and sifted to meet the ¼-in. limitation. These products are used in lightweight concrete. The significant weight reduction is, by comparison, for lightweight concrete, 30–90 lb/ft^3 (pounds per cubic foot) * and for normal-weight concrete, 90–110 lb/ft^3.

Look again at Figure 1–1 and notice that coarse aggregates account for 31 to 51 percent of the bulk of concrete. Coarse aggregates are particles whose sizes exceed ¼ in. The most common coarse aggregate is gravel; however, crushed stone and air-cooled blast-furnace slag may be used for normal concrete. Pumice, scoria, perlite, vermiculite, diatomite, shale, and clay are used in lightweight concrete.

---

* The notation lb/ft^3 means the same as pounds per cubic foot where the / means "per."

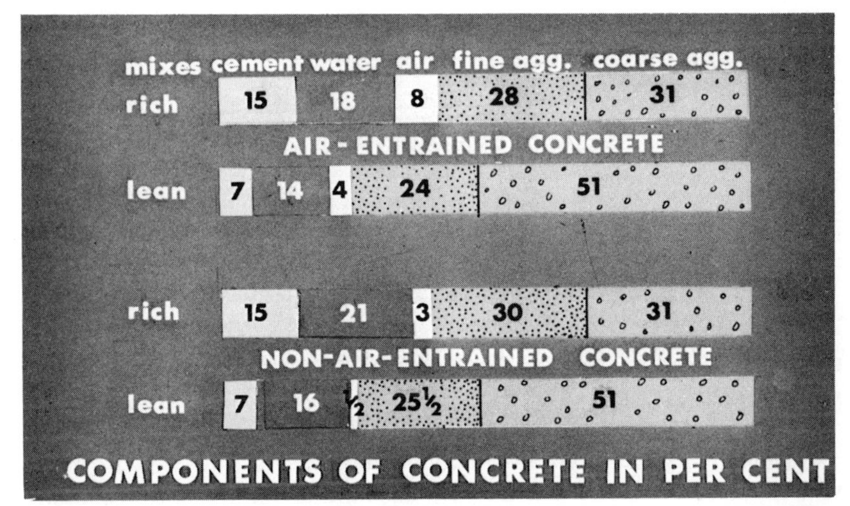

Figure 1–1 Proportions of Concrete Mixtures *(Courtesy of Portland Cement Association)*

The shape of coarse aggregate should be similar to that of fine aggregate: round or cubical. Elongated particles, those made from shaly or porous particles, must be avoided because they create problems, especially with regard to weathering.

Aggregates, fine and coarse, constitute 60 to 80 percent of the bulk of concrete. They also are capable of containing water, the agent that combines with cement to make a paste. Figure 1–1 shows that water accounts for 14 to 18 percent of the bulk of a concrete mixture. These percentages of water include water added to a mixture, water content in the aggregate, and water surrounding the aggregates. Any aggregate may have a varying water content: from oven dry, with none; some, because of air drying; damp; and saturated (Figure 1–2). This factor is extremely important because it must be accounted for in concrete mixing. The amount of tap water added to a mixture is controlled, judged, or measured by the method of making one or several trial batches. Generally, if bone-dry aggregates are used, greater quantities of tap water are needed; less tap water is needed where damp or wet aggregates are used.

Since aggregates have the ability to retain water, they will absorb water from the paste. This, in turn, causes some separation of paste and poor-quality concrete. In addition, freezing and thawing temperatures expand and contract the water, causing expansion and contraction of the concrete, resulting in structural damage to the mass.

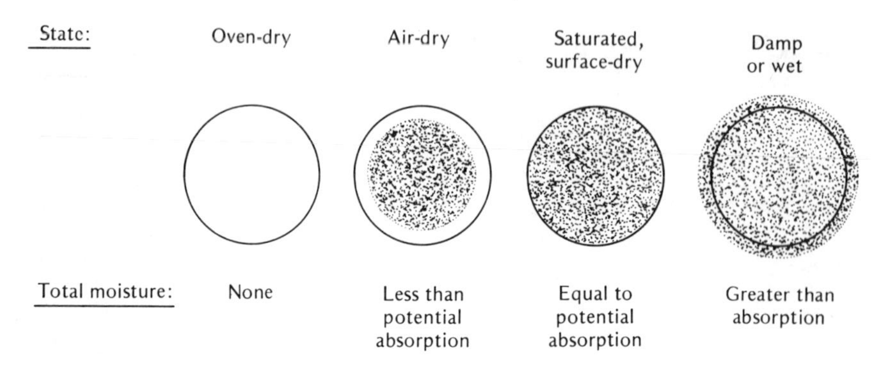

Figure 1–2 Water Content in Aggregate *(Courtesy of Portland Cement Association)*

Sufficient water should be added to create the desired mixture, whether dry mix or normal mix, by accounting for the water content percentage in the aggregate and maintaining the water/cement ratio needed for high-quality concrete.

Cement, more properly called *portland cement,* is a manufactured product that when mixed with water creates a paste. This product was invented by an English mason, Joseph Asphin. Portland cement consists of portions of lime, silica, alumina, and iron components. An involved process of preparation is used that creates a portland cement clinker. This clinker is then pulverized to a fine powder and sold as cement.

Different types of cement are manufactured for a variety of uses. These are listed in Table 1–1.

Air-entraining additive or *admixture* is a recent development in the production of concrete. All concrete has natural air-entraining properties. These are trapped air pockets within the concrete. As Figure 1–1 shows, the additive may range from 4 to 8 percent, and by its inclusion provides a very desirable quality that resists weathering.

Consider that the freezing and thawing of concrete cause expansion and contraction of the concrete. With little or no air entraining, the structure is exposed to severe pressures. However, with air entraining the mass contains from 300 to 500 billion bubbles per cubic yard of concrete. The variation of temperatures and weather cause the same conditions, but the bubbles allow for expansion and contraction without disturbing the mass. In addition, greater workability is obtained and a better cement/water ratio may be used, again improving the quality of the concrete.

Coloring agents are natural, such as aggregates of quartz, marble, granite, or ceramic, or they may be pigments, such as pure mineral

Table 1–1 Cement Types and Uses

Type	*Uses*
ASTM Type 1, CSA * normal (*Note:* ASTM Types 2, 3, 4, and 5 are used for commercial applications having special requirements.)	General-purpose cement, suitable for all uses except when special requirements exist: pavements, sidewalks, reinforced concrete, bridges, tanks, reservoirs, and masonry units.
White portland cement	True portland cement that is white instead of the customary gray. It is used for architectural purposes: e.g., curtain wall, terrazzo, stucco, cement "paint," tile grout, and decorative concrete.
Portland blast-furnace slag cement	Made from blast-furnace slag with and without air-entraining additive. Used where normal cement is used and for general concrete construction.
Masonry cements	Cements that conform to a mixture of portland cement, air-entraining additives, and supplemental materials selected for their ability to impart workability, plasticity, and water retention to masonry mortars. Used in laying brick and block and for other mortar needs.
Waterproof cement	Portland cement made by adding a small amount of calcium, aluminum, or other stearate. Adds to the waterproofing quality and characteristics of the cement.
Plastic cements	Cements that have plasticizing agents added to Type 1, normal, and Type 2 cement during manufacture. This cement is commonly used for making mortar, plaster, and stucco.

* The types of cement listed are selected from a complete list of portland cements as published in *Design and Control of Concrete Mixtures*, Portland Cement Association, 1968.

oxides that are ground finer than cement, with colors ranging from white through the color spectrum to black.

Where natural coloring agents are used, concrete receives a special finishing method called *exposing*. A washing away of surface cement is done to expose the natural color of the aggregate.

Where oxides are used, the powdered form is added to the batch or applied atop the surface. In either event the amount used must not exceed 10 percent of the weight of the cement. For reference and example: 1½ lb of pigment per 100 lb of cement produces a pastel color, whereas 7 lb of pigment per 100 lb of cement produces a deep color.

The *strength* of concrete is measured in laboratory environments using samples of concrete and special pressure machines. Figure 1–3 shows the final analysis of the tests. Clearly, the strength of normal concrete is hardest after 28 days of curing and further note that the water/cement ratio has a significant bearing on its strength. The lower the water content with respect to the cement, the greater the strength.

Two factors play a significant part in determining strength: temperature and curing method. Cool temperatures are ideal for creating concrete's strength because they allow slow evaporation and ample

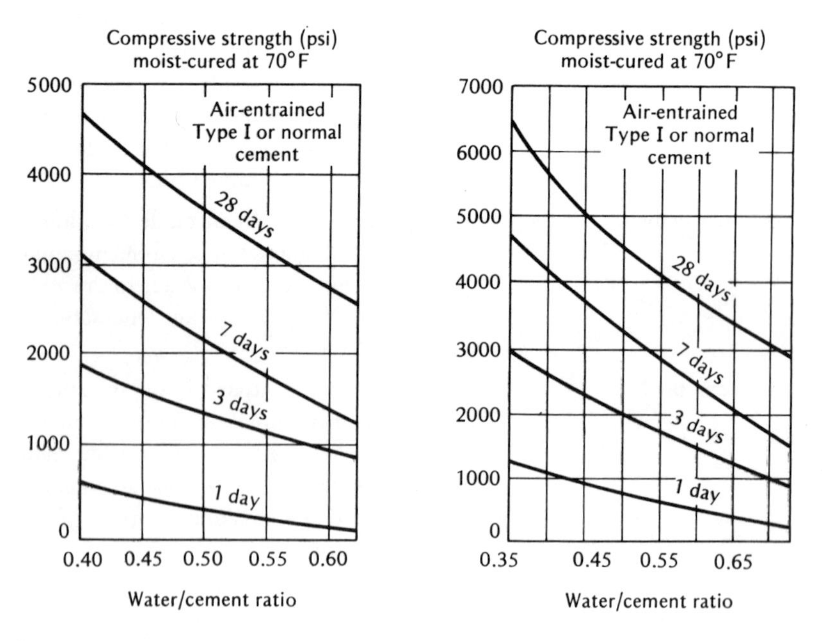

Figure 1–3 Normal Concrete's Strength *(Courtesy of Portland Cement Association)*

time for curing. Hot temperatures have the most adverse effects, as they accelerate evaporation; generally more water is needed, but if the water/cement ratio is to be maintained, more cement is used. This acceleration causes the mixture to harden rapidly, preventing or greatly reducing the curing time needed to establish the strength. As a rule, concrete should be poured on days where the temperature is above 32°F and below 90°F.

Weight is a physical property of concrete, and there are in general four categories:

1. Thermal-insulating lightweight concrete, with unit weights ranging from 15 to 90 lb/ft^3.
2. Lightweight concrete, with a unit weight of less than 115 lb/ft^3.
3. Normal concrete, with unit weights ranging from 90 to 180 lb/ft^3.
4. Heavyweight concrete, with unit weights ranging to 400 lb/ft^3.

TABLE 1–2 CONCRETE PRESSURES FOR COLUMN AND WALL FORMS (COURTESY OF AMERICAN PLYWOOD ASSOCIATION)

Pour Rate	*Pressures of Vibrated Concrete (psf) (a), (b), (c)*			
	50° F		*70° F*	
(ft./hr.)	*Columns*	*Walls*	*Columns*	*Walls*
1	330	330	280	280
2	510	510	410	410
3	690	690	540	540
4	870	870	660	660
5	1050	1050	790	790
6	1230	1230	920	920
7	1410	1410	1050	1050
8	1590	1470	1180	1090
9	1770	1520	1310	1130
10	1950	1580	1440	1170

Notes: (a) Maximum pressure need not exceed 150h, where h is maximum height of pour.
(b) For non-vibrated concrete, pressures may be reduced 10%.
(c) Based on concrete with density of 150 pcf and 4 in. slump.

TABLE 1–3 CONCRETE PRESSURES FOR SLAB FORMS (COURTESY OF AMERICAN PLYWOOD ASSOCIATION)

Depth of Slab (in.)	*CONCRETE PRESSURE (psf)*	
	Non-Motorized Buggies (a)	*Motorized Buggies (b)*
4	100	125
5	113	138
6	125	150
7	138	163
8	150	175
9	163	188
10	175	200

Notes: (a) Includes 50 psf load for workmen, equipment, impact, etc.
(b) Includes 75 psf load for workmen, equipment, impact, etc.
(c) Forms for concrete slabs must support workmen and equipment as well as the concrete. Table 1–3 gives form pressures which represent average practice when either motorized or non-motorized buggies are used for placing concrete. These pressures include the effects of concrete, buggies, and workmen.

Force is also a physical property of concrete that must be considered in its liquid or plasticized state and when hardened. In its liquid state, concrete exerts a lateral force in pounds per square foot. Tables 1–2 and 1–3 indicate that a significant force exists against a form and that the rate of pouring the concrete bears directly on the amount of force (pressure). The hardened concrete exerts a dead-weight force, usually downward, at a rate of its total weight, as in a wall, or column or distributive weight, as in a slab.

Thermal conductivity has become a very important factor recently because of the soaring costs of fuel for heating and cooling. Normal-weight concrete has the property of allowing heat and cold to penetrate. For instance, where slabs are placed directly on the ground, radiation from the ground frequently affects the measured temperature on the exposed surface. This is also possible on vertical walls made from the same materials.

When lightweight concrete is made using, for instance, vermiculite as a coarse aggregate, the concrete takes on a *thermal-insulation* quality. This feature aids significantly in the control of inside environmental temperatures. As heat is felt by the vermiculite particles, they react by expanding, thereby creating a barrier. As the heat intensifies, more expansion occurs, until maximum thermal protection is exceeded. At that time heat penetrates into the interior of the structure. Lightweight concrete panels are frequently used in roofing and side walls.

By comparison, the thermal conductivity of normal concrete will range from 9 to 12 Btu per hour per inch of thickness per square foot per degree F; whereas lightweight concrete's thermal conductivity will range from less than 1 to 4 Btu/hr/in. thickness/ft^2/°F.

CHEMICAL PROPERTIES

Concrete is made possible because of the chemical reaction of cement and water. This reaction, called *hydration,* requires time and favorable conditions of temperature and moisture.

Water is required initially to begin the process by making a cement paste. Once the compound is formed, hydration starts and con-

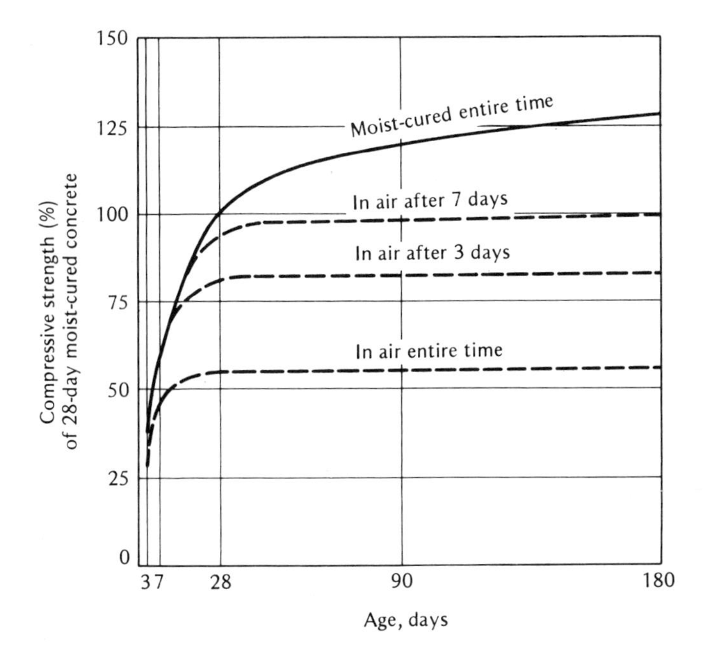

Figure 1–4 Hydration at Work *(Courtesy of Portland Cement Association)*

tinues as long as moisture remains. The process stops when moisture is removed by heat and evaporation, but it can be restarted by adding moisture, even though this is undesirable.

The chart in Figure 1–4 shows the strength that is formed under various conditions of hydration. When the hydration process is controlled for the first 28 days, that is, when moisture is continually available in the mass, 100 percent strength is obtained. If hydration is continued beyond that point, strength exceeds 125 percent if controlled for 180 days. However, notice that various degrees of strength are obtained if improper methods of curing are used.

Chemical action can be sustained by:

1. Keeping the concrete wet 24 hr/day, using water and hoses.
2. Applying a plastic cover over the concrete, trapping the moisture within.
3. Covering the concrete with burlap, which is kept wet.
4. Covering the concrete with fresh hay, which is kept wet. This method also aids against freezing.

TYPES OF MIXES

Each mixture must have three basic objectives:

1. Workability of freshly mixed concrete.
2. Required properties of hardened concrete.
3. Economy.

To achieve workability, proportioning of the materials of concrete must result in a mixture that moves as needed. Required properties of hardened concrete, such as freeze–thaw resistance, water tightness, wear resistance, and strength, are dependent on suitable low water/cement ratios as well as entrained air. Economy can be achieved to minimize water and cement by using the stiffest mixture practical, the largest practicable maximum size of aggregate, and the optimum fine aggregate/coarse aggregate ratio.

Water/cement ratios should be maintained at 49 to 53 percent for all general concrete uses. These include railings, curbs, sills, ledgers, slabs, walks, retaining walls, and even underground slabs, such as cellar walls, that come in contact with earth.

TABLE 1–4 PROPORTIONS BY WEIGHT TO MAKE 1 CUBIC FOOT OF CONCRETE AND METRIC CONVERSION *

MAXIMUM SIZE	*AIR-ENTRAINED CONCRETE †*				*CONCRETE WITHOUT AIR*			
Coarse Aggregate (in.)	*Cement (lb)*	*Fine Aggre-gate (lb)*	*Coarse Aggre-gate (lb) ‡*	*Water (lb)*	*Cement (lb)*	*Sand (lb)*	*Coarse Aggre-gate (lb) ‡*	*Water (lb)*
⅜	29	53	46	10	29	59	46	11
½	27	46	55	10	27	53	55	11
¾	25	42	65	10	25	47	65	10
1	24	39	70	9	24	45	70	10
1½	23	38	75	9	23	43	75	9

BY METRIC PROPORTIONS TO MAKE APPROXIMATELY 1 CUBIC FOOT * OF CONCRETE

Coarse Aggregate (cm.)	*Cement (kg)*	*Fine Aggre-gate (kg)*	*Coarse Aggre-gate (kg)*	*Water (kg)*	*Cement (kg)*	*Fine Aggre-gate (kg)*	*Coarse Aggre-gate (kg)*	*Water (kg)*
0.9	13.2	24	21	4.5	13.3	27	21	5
1.5	12.3	21	25	4.5	12.3	24	25	5
1.875	11.4	19	30	4.5	11.5	21	30	4.5
2.5	11.0	18	32	4	11.0	20	32	4.5
4.0	10.5	17	34	4	10.5	19	34	4

* One twentieth of a cubic meter is approximately equal to one twenty-seventh of a cubic yard, or 1 cubic foot.
† Air-entrained concrete must be mixed in a machine mixer for proper distribution of admixture.
‡ If crushed stone is used, decrease coarse aggregate by 3 lb and increase sand by 3 lb.

TABLE 1–5 PROPORTIONS BY VOLUME

MAXIMUM SIZE	AIR-ENTRAINED CONCRETE				CONCRETE WITHOUT AIR			
Coarse Aggregate (in.)	*Cement*	*Sand*	*Coarse Aggregate*	*Water*	*Cement*	*Sand*	*Coarse Aggregate*	*Water*
3/8	1	2 1/4	1 1/2	1/2	1	2 1/2	1 1/2	1/2
1/2	1	2 1/4	2	1/2	1	2 1/2	2	1/2
3/4	1	2 1/4	2 1/2	1/2	1	2 1/2	2 1/2	1/2
1	1	2 1/4	2 3/4	1/2	1	2 1/2	2 3/4	1/2
1 1/2	1	2 1/4	3	1/2	1	2 1/2	3	1/2
1 1/2	1	2 1/2	4	1/2	1	3	4	1/2

The types of mixes suited to the various projects listed in the previous paragraphs are as follows:

Mix	*Project*
1:2:2¼ (max. ¾-in. aggregate)	Concrete subject to extreme wear and weather, driveways
1:2¼:3 (max. 1-in. aggregate)	General slabs, sidewalks, driveways, patios, floors, retaining walls
1:3:4 (max. 1½-in. aggregate)	Foundations, footings, and walls

As a general guide, when specifying coarse aggregate size, limit the maximum size to no more than one third of the slab or wall thickness (e.g., 4-in. slab, 1–1¼-in. maximum aggregate size).

Tables 1–4 and 1–5 provide data for mixing small batches of concrete using different sizes of coarse aggregate. The results of using these tables provide a mixture of approximately 1:2¼:3 and a water/cement ratio of 49 to 50 percent. The water quantity is based upon a condition that the sand and coarse aggregate are saturated (wet), that no water will be absorbed by these elements, and that all water added to make concrete is used to make paste. Table 1–4 lists quantities of ingredients needed for 1 cubic foot (1 ft^3) of concrete. If 2 cubic feet are needed, double all ingredient weights.

In contrast, small jobs can be accomplished by using premixed sacks of concrete ingredients, and these are sold in most lumberyards and builder's supply stores. Carefully follow the mixing instructions written on the sack. Water must be carefully measured so that the water/cement ratio is maintained and the slump is controlled. A 90-lb package makes ⅔ ft^3 of concrete.

Methods of Mixing

Of the two methods of proportioning shown in the tables, the proportion by weight is the better. To accomplish this method, an ordinary bathroom scale and bucket are needed.

Weigh the bucket empty or adjust the scale to zero with the empty bucket resting on the scale. Then measure out the proportions and dump them into a mixing container, box, board, wheelbarrow, or machine mixer (Figure 1–5).

If a small 3-cubic foot (3 ft^3) machine mixer is used, multiply each element used by a factor of 3; for example, ⅜-in. aggregate,

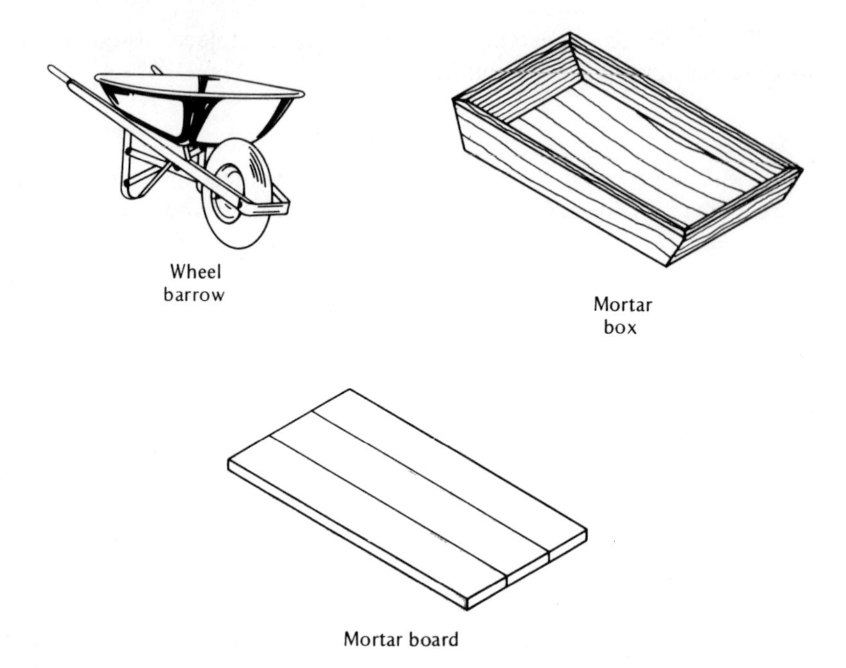

Figure 1–5 Types of Mixing Vessels

3 × 46 lb = 138 lb; sand, 3 × 56 lb = 168 lb; cement, 3 × 29 lb = 87 lb; and water, 3 × 10 lb = 30 lb. Water volume to the pound: 1 gal, 7.6 lb.

The bucket can be used to measure ingredients for small batches by volume, although this method is not as reliable as the weighing method.

If batches approximating one cubic yard or more are needed for a project, mixing by hand or with a small machine mixer requires a great deal of manual labor. The work involved equates to moving more than 1 ton of material. The alternative is to order ready-mixed concrete from a local company.

The customary hand method for mixing a batch of concrete is to thoroughly mix all dry ingredients, which have been previously carefully measured. When a gray color is even throughout the mixture, it is ready for the water. Part the mixture to form two halves in a box or, if on a flat board, a donut. Pour the measured water into the space and with a hoe draw the dry ingredients into the water, mixing well.

If a machine mixer is to be used, the following sequence should be followed. With the mixer stopped, add all the coarse aggregate and

Figure 1–6 Various Mixture Problems of Concrete *(Courtesy of Portland Cement Association)*

half of the mixing water. If an air-entraining agent is used, mix it with this part of the water. Start the mixer; add the sand, cement, and remaining water; and allow the mixture to blend for at least 3 minutes. Dump the concrete from the mixer and cleanse the mixer with clean water.

Except for premixed bags, all concrete work should be started by making a trial batch. A trial batch could be a 1-ft^3 batch or a percentage, say 10 percent, of a 1-ft^3 batch. When mixed, it should be workable, not too wet, too stiff, too sandy, or too stony, as shown in Figure 1–6.

QUESTIONS

1. What is a major objective in the study of masonry?
2. Compare fine aggregates with coarse aggregates.
3. What type of cement is used in brick and block laying and other mortar needs?
4. True or false: The lower the water content with respect to the cement, the greater the strength.
5. What types of conditions should exist before concrete is poured?
6. Discuss hydration.
7. Which method of proportioning is the most effective?

Chapter 2

Footings

Dry mix a mixture of concrete in which the water content is greatly reduced.
Elevation plan a part of a blueprint that shows details in a vertical (up-and-down) view.
Footing a base for a wall or other structure that provides stability for that structure.
Form a parameter or set of parameters made from earth or wood and, on occasion, steel that contains the footing concrete.
Foundation plan a cross-sectional view of a building's foundation as viewed from above, containing all dimensional detail.
Grade line a point of relative height set either in specifications or on elevation plans.
Keying a method of using a key in a form to create a depression to permit bonding of two separate pours of concrete.
Pier a free-standing column.
Pilaster a projection from a masonry wall that provides strength for the wall and a bearing surface for lateral beams.
Specifications details relative to building that are not included in plans but are vital to the building's construction.

OBJECTIVES—INTRODUCTION

The first critical phase of a construction job is the location, forming, and pouring of the *footing*. This chapter develops the need for the footing, numerous types of footings, many ways to form footings, as well as steel and concrete requirements. The chapter is finalized with a quality-control checklist.

All the data above can be phrased as two specific objectives: *to be able to translate a footing blueprint into the footing itself* and *to understand the need for and purpose of footings.* An architect will detail the footing requirements in the foundation and elevation plans, including top and side views. The ability to recognize and understand these drawings is essential.

Consider what your legs would be like without feet. Could you maintain balance? Would you sink into soft earth? A footing is to a wall or other structure as your foot is to your legs and body.

READING AND UNDERSTANDING THE FOOTING BLUEPRINT AND SPECIFICATION DATA

As mentioned above, footing details are to be found in the foundation and elevation drawings within a set of plans. Figure 2–1 shows a segment of a typical foundation plan for a walled building, and Figures

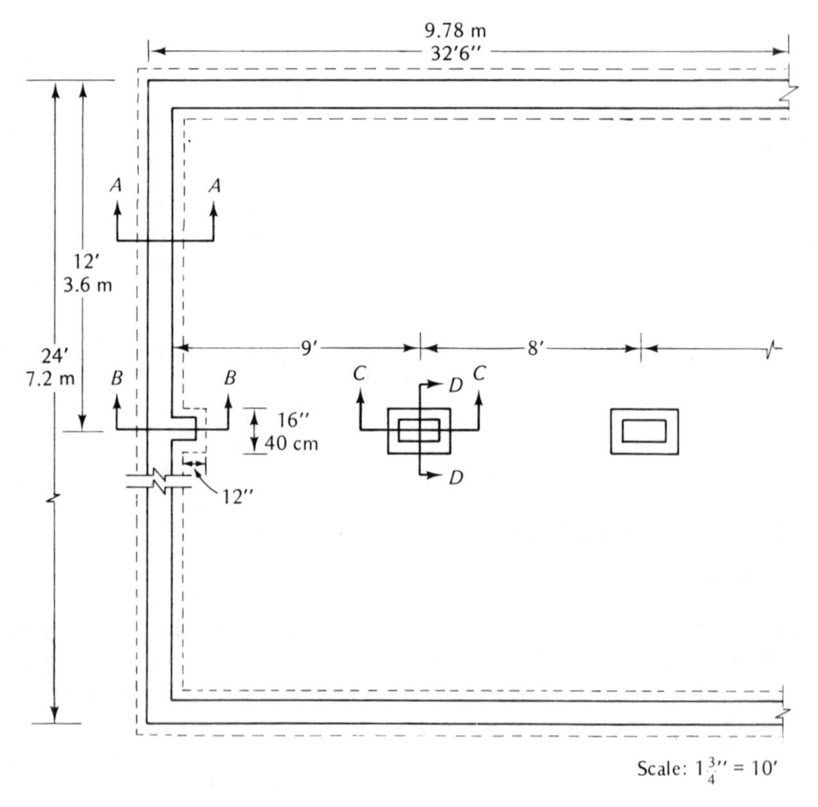

Figure 2–1 Foundation Plans

2–2 through 2–5 show the corresponding elevation details of the footings for the walls, piers, and pilaster. Each element in the drawings is examined, using both figures where applicable.

Common Wall Footings

Look at the left side of Figure 2–1 and find the detail designation AA. This is an identity indication that an elevation drawing contains the specific data on the *common wall footing* and wall. Now look at Figure 2–2. It contains the details (elevation) concerning the footing for the common wall that are needed by the builder. In this detail drawing, the useful data are as follows:

- **a.** Footing height—8 in. or 20 cm.[1]
- **b.** Footing width—16 in. or 40 cm.
- **c.** Wall thickness—8 in. or 20 cm.
- **d.** Footing overhang on each side of the wall—4 in. or 10 cm.
- **e.** Steel rods placed in the footing for reinforcement.

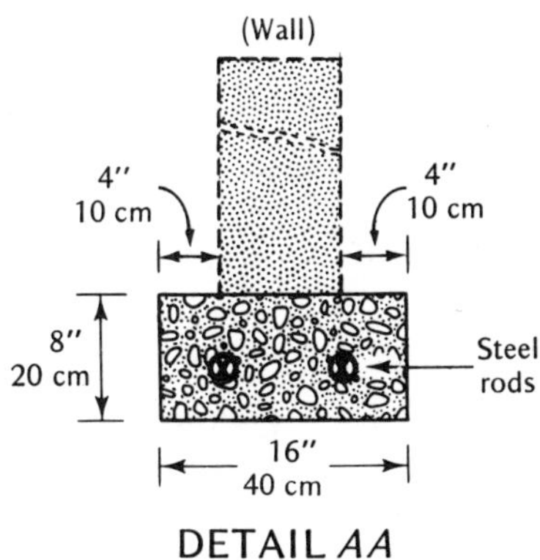

Figure 2–2 Common Wall Footings

Using these data, the first specification for the job can be listed: "Common wall footings shall be 8 in. high by 16 in. wide made of poured concrete with two ½-in. reinforcement rods" (in metric, 20 cm replaces 8 in. and 40 cm replaces 16 in.).

---

[1] Centimeter is written as "cm" in combined form.

Pilaster Footings

Look again at Figure 2–1 and see that detail designation BB concerns the wall and pilaster. Figure 2–3 shows the elevation parameters of the footing, wall, and pilaster. In this detailed drawing, the following data are provided:

a. Common footing dimensions, including two steel rods.
b. Extended footing needs for the pilaster section, which also allows for a 4-in. or 10-cm overhang beyond the pilaster.
c. To reinforce the pilaster form, additional steel rods are used, lapping the wall footing rods at 90 degrees.
d. The pilaster portion of the footing would be poured simultaneously with the wall footings.

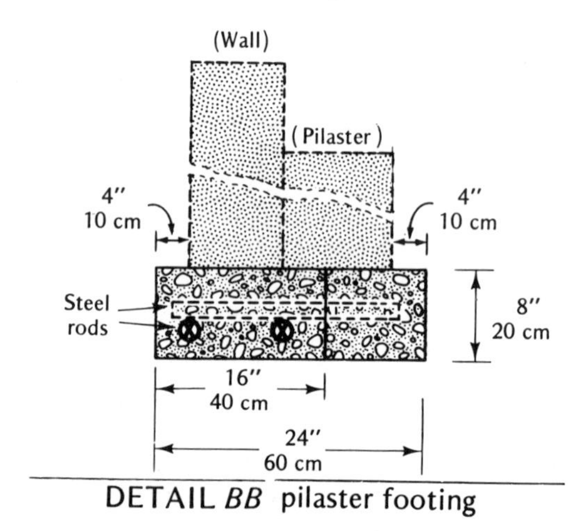

Figure 2–3 Elevation Footings of Pilaster Footing

The second specification should include reference to the first one and added data: "Make pilaster footing depth and height equal to common wall footing; include steel in the pilaster portion of the footing and make for a continuous pour."

Pier Footings

Return again to Figure 2–1 and locate the pier detail, designated by CC and DD. The pier supports the center girder of the floor assembly.

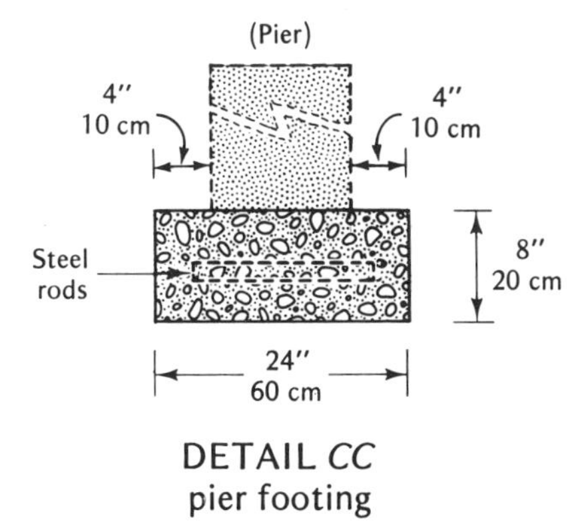

Figure 2–4 Pier Footing Length

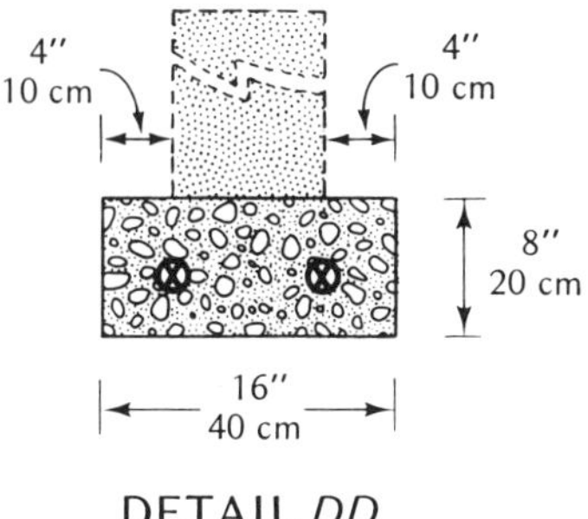

Figure 2–5 Pier Footing Width

Details about CC and DD are formed in Figure 2–4, pier footing length, and Figure 2–5, pier footing width. Notice that the pier footings carry the same 10-cm or 4-in. expanse greater than the pier itself. Also, the pier's thickness is the same as the common wall footing. The figures show that there are two steel reinforcing rods in each footing. If desired, a single rod shaped like a square donut can be fashioned and used.

These footings carry the same specification data as the common wall, so rather than have a series of specifications for each part of the footing, one will usually do. "Footings shall conform to plans for depth, width, and length, shall have two ½-in. steel rods placed within the concrete, and the concrete shall use aggregate no larger than 1½ in. The footing shall be, where possible, a continuous pour."

Grade Marks

Footings shown on elevation plans include a grade reference point. This point, as Figure 2–6 shows, is usually determined from the center/crown of the street. The footing top surface is usually given, and from this point either the form top edge or, if earth forms are used, a grade mark stake is driven into the footing area.

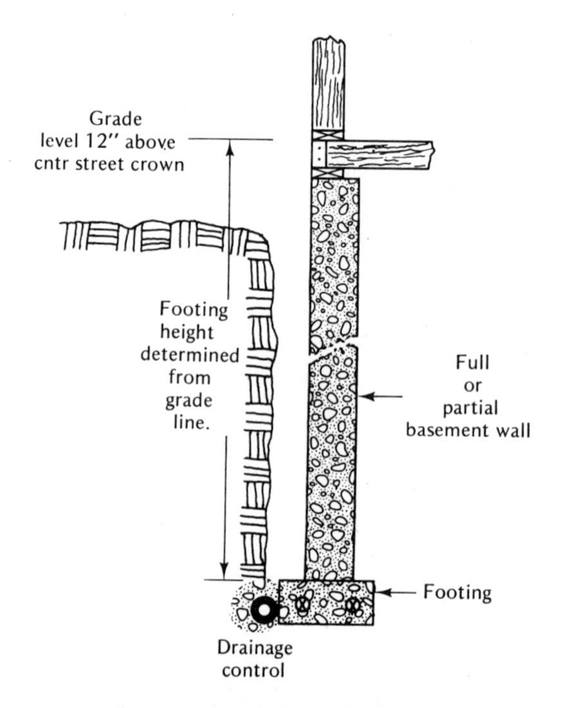

Figure 2–6 Grade Reference Point

Retaining-Wall Footings

Retaining walls are often needed where homes are built on uneven terrain. They may be a part of the foundation or separate, as, for instance, to support an embankment of earth. These walls need footings similar to the common wall, especially with regard to width and thickness. They should extend at least 10 cm or 4 in. beyond the wall surface if perpendicular as Figure 2–7 shows, or as the plumb bob shows in Figure 2–8.

Figure 2–7 Vertical Retaining-Wall Footing

A
(Wall)
2′ to 4′
60 to 120 cm
8″
20 cm
Steel
rods
B
Retaining-wall footing

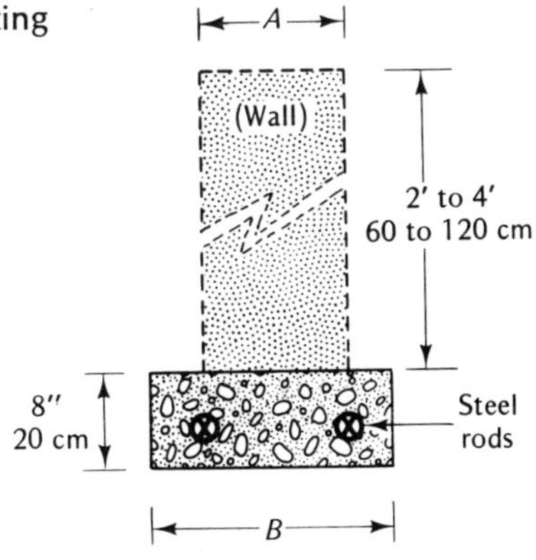

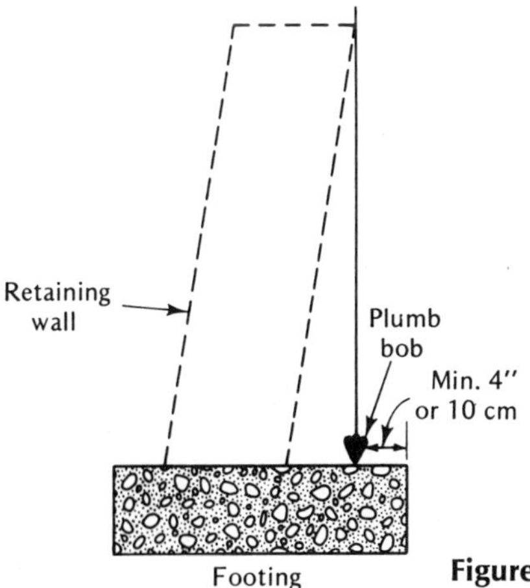

Figure 2–8 Slanted Retaining-Wall Footing

Soil and the Footing

All varieties of soil may be encountered when preparing for footings. All firmly packed soils containing sand, gravel, stone, and dirt require the same type of footing. However, clay is a poor footing base. It is subject to movement, and when wet will allow footings and walls to sink and shift. Where clay is found, it should be removed if possible and suitable fill installed, or a wider footing can be used. See the local building code.

Steel

The addition of steel rods in a footing makes it qualify as a reinforced concrete footing. The rods provide tensile strength. The bars used for reinforcement are deliberately made with a rough texture, as Figure 2–9 shows. Tests have shown that rough surfaces provide better bonding.

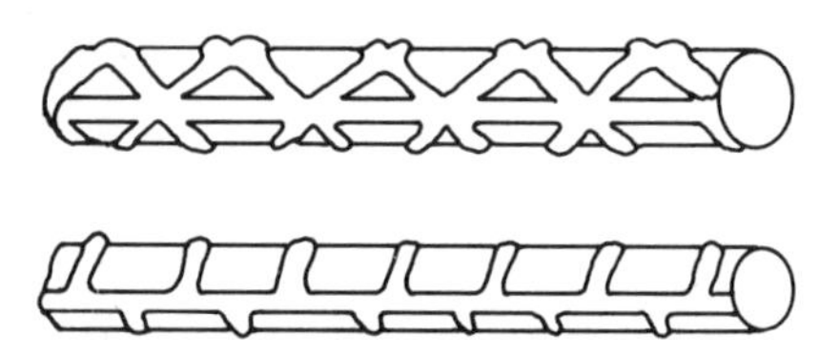

Figure 2–9 Bars Used for Reinforcement

Each of the detail drawings indicates reinforcement rods in the footing. These rods are available in sizes ranging from ⅜ to 2¼ in. in diameter. For light and standard footings as well as general concrete wall structures, rods ⅜ to ¾ in. are usually used. The specifications will dictate the rod size, by *bar designation number*, as follows:

BAR NO.	DIAMETER (IN.)	CM *
3	⅜ (0.375)	0.9
4	½ (0.500)	1.27
5	⅝ (0.625)	1.59
6	¾ (0.750)	1.9
7	⅞ (0.875)	2.2
8	1 (1.000)	2.54

* Approximate.

Lapping bars within the form area is standard procedure. The laps may be wired or left adjacent.

Quality and Quantity of Concrete

Footings generally require a concrete quality of 2500 to 3500 psi compression in 28 days. Also, concrete may be mixed to a workability having as much as a 4-in. slump. Use Table 2–1 to aid in making esti-

TABLE 2–1 FOOTING REQUIREMENTS FOR RETAINING WALLS *

Wall Thickness	*Footing Width*	*Footing Thickness*	*Steel Rods*	*Concrete for Footing (linear measures)*
6 in.	14 in.	8 in.	1/2 in.	1/3 ft^3
15 cm	35 cm	20 cm	1.2 cm	0.07 m^3
8 in.	16 in.	8 in.	1/2 in.	1/2 ft^3
20 cm	40 cm	20 cm	1.2 cm	0.08 m^3
12 in.	20 in.	8 in.	1/2 in.	2/3 ft^3
30 cm	50 cm	20 cm	1.2 cm	0.1 m^3
16 in.	24 in.	8 in.	1/2 in.	9/10 ft^3
40 cm	60 cm	20 cm	1.2 cm	0.12 m^3

* 27 ft^3 = 1 yd^3.

mates of concrete needed for footings. Values are provided in both inches and metric systems. Find the wall thickness from plans and compare or establish footing width and thickness using the second and third columns. Then order the amount of concrete needed by calculating the number of linear feet or meters times the factor given in the last column.

THEORY OF FORMING THE FOOTING

Several methods are used to form footing areas. Each serves the primary purpose of confining to a specific area the concrete needed for the footing.

Earth Form

The earth-form method shown in Figure 2–10 may be used in firmly packed earth and sand. It is very difficult to use in rocky soil. The object is to shape the area of the footing to its precise width and depth. Although not shown in the figure, grades stakes indicating the top surface level of the footing would be driven periodically along the formed area. After the concrete is poured, the stakes are removed.

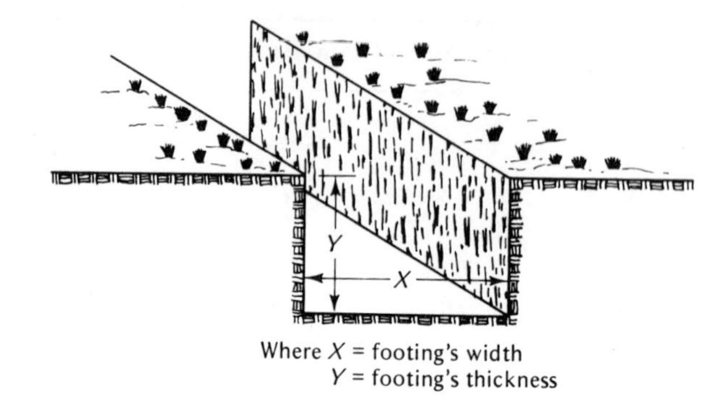

Figure 2–10 Earth-Form Method

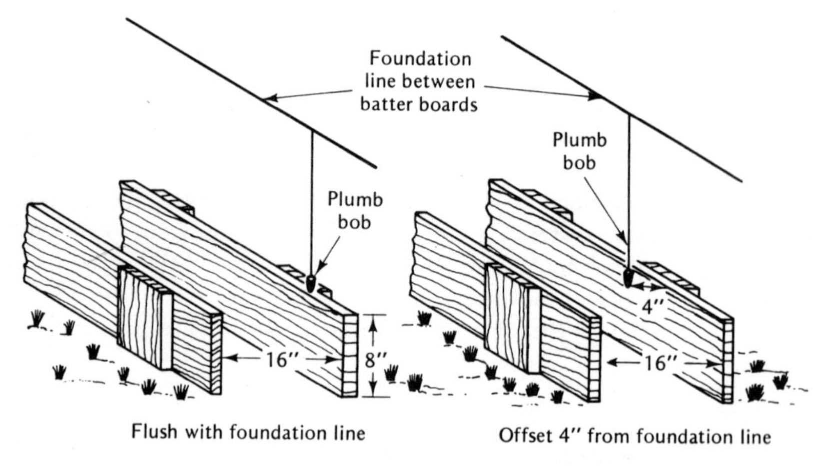

Figure 2–11 Wood Footings

Lumber Forms

Footings need to be formed with lumber wherever the earth is not adequate to form sides. One- and 2-in. lumber stock are commonly used. Figure 2–11 shows a wood footing. Carefully note that the footing is staked on its outside and also that one or both edges of the footing are placed according to the foundation line.[2] The 4-in. offset corresponds to

---

[2] All carpenter tasks related to forming may be learned from *Carpentry for Residential Construction* by Byron W. Maguire (Reston, Va.: Reston Publishing Company, 1975).

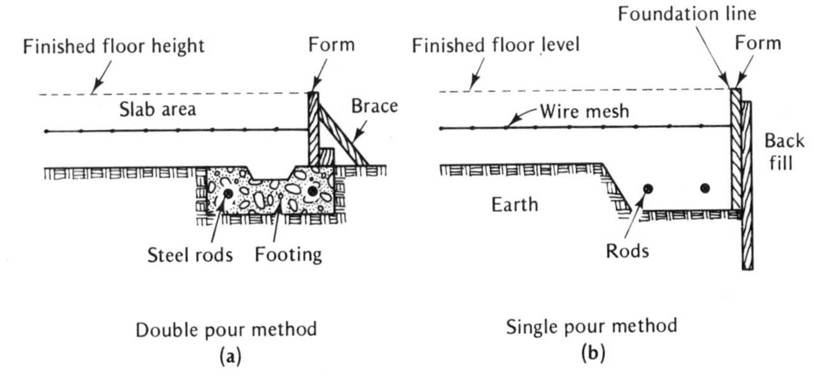

Figure 2–12 Slab Footings

the drawing used in the specifications segment of this chapter. The flush is used for a slab footing, as Figure 2–12 shows.

Slab Footings The footing for a slab may be poured separately (Figure 2–12a) or as part of the single pour (Figure 2–12b). Method A uses the earth or board method previously examined; but method B indicates a board only on the outside edge, with earth forming the inner surface.

Keyed Footing Notice that Figure 2–12a shows a footing with a depression. This is called a *key*. Figure 2–13 defines this forming method more clearly. A nominal 2 × 4 is tapered so that it is easily removed once the concrete footing has set. The figure shows how the final product looks.

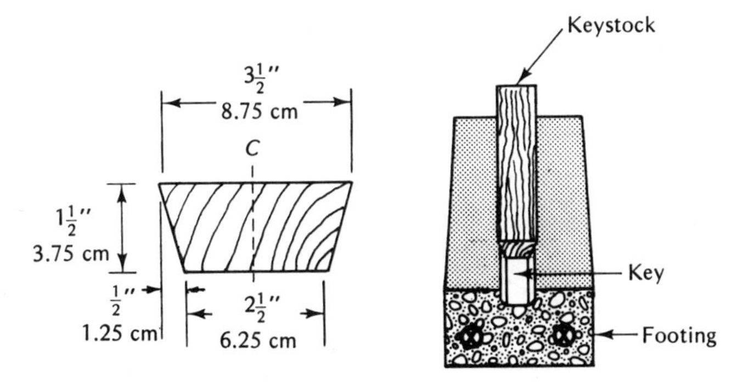

Figure 2–13 Defined Forming Method

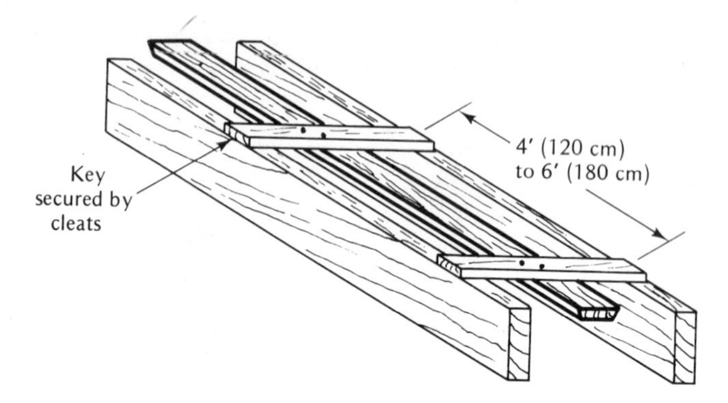

Figure 2–14 What Form Must Be Like if Keying Is Used

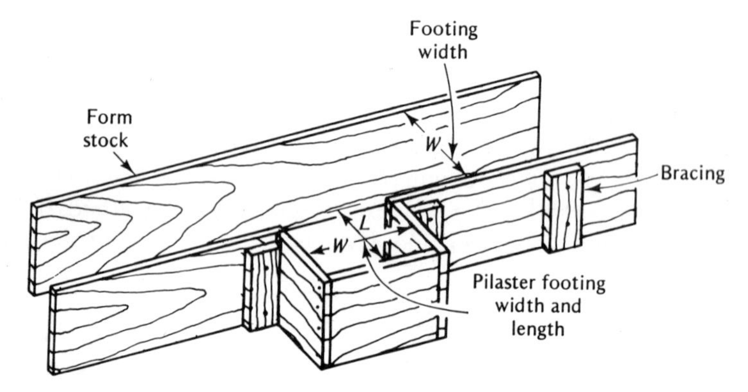

Figure 2–15 Pilaster Form

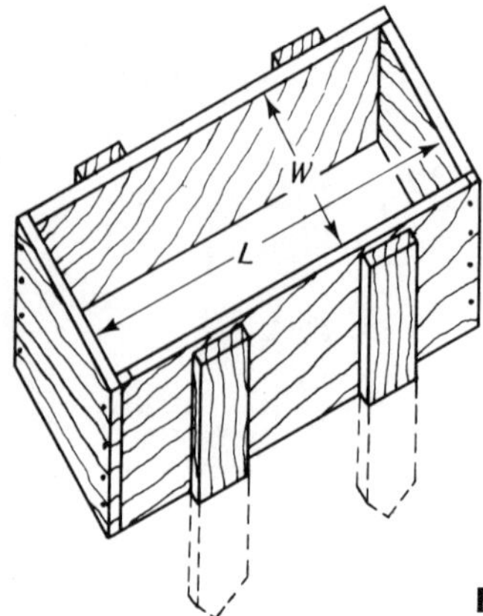

Figure 2–16 Pier Form

Figure 2–14 shows what the form must look like if keying is used. Nominally 1 × 2 or 1 × 3 slats are used to support the key stock within the form area.

Pilaster and Pier Forms The forms discussed thus far are wall and slab forms. Pilaster forms are a modification of the wall form (Figure 2–15), where their dimensions cause a projection along the wall. Pier footings are small sections of wall forms having both sides and ends (Figure 2–16).

THEORY OF POURING AND FINISHING FOOTINGS

The first step in the pouring phase should be setting the steel rods within the form. The rods should be fairly well centered top to bottom and placed side to side so that there are several inches of concrete between rod and form. The rods may be suspended from the form by wire ties or blocked in position with brick or stone. An alternative method of setting rods is to fill the footing one-half full with concrete, set the rods, and pour the remaining concrete. Of course, this method is impractical if keying is used.

Pouring

Following the setting of the steel, the concrete is poured into the form. There are two types of mixtures used: *standard mix* and *dry mix.*

Standard mix contains all the water required, and as it is poured into the form, rakes, short lengths of 2 × 4s, or a mechanical vibrator should be used to tamp the mixture well into the form and around the steel rods.

Dry mix does not contain the water required for workability. It is used where the surfaces of and around the form contain excessive amounts of natural water. This natural water will mix with the dry mix, bringing the final mixture to the needed water content. The dry mix is poured into the form and tamped just as described for the standard mix.

Screeding the footing mixture is done concurrently with filling and tamping. This leveling of the surface may be done with the back of a rake or by use of a screed board. The object is to level and smooth the concrete top surface to the parameters of the form's top edges, and grade level.

Curing

The mixture must be kept moist for 7 days to achieve a 2500-psi compressive strength (3500 psi after 28 days). Water may be sprayed directly on the *set* concrete or wetted burlap bags may be used.

INSPECTION OF THE FOOTING

Inspection	*Satisfactory/N.A.*	*Unsatisfactory*
Has the footing been properly located?	______	______
Have the specifications been studied and applied?	______	______
Do earth walls and base conform as a form?	______	______
Is soil adequate to support the footing and wall?	______	______
Are wooden forms installed properly so that		
a. Bracing is proper?	______	______
b. Variations in dimensions are ±½ to 2 in. only?	______	______
c. Reduction in thickness is less than 5 percent?	______	______
d. Variation of building lines are only ¼ to ½ in. in 40 ft?	______	______
Are steel rods of correct size and placed properly?	______	______
Is key stock installed?	______	______
Has the proper type of mixture been ordered and delivered?	______	______
Has the pour been tamped, leveled and screeded?	______	______
Have curing steps been followed for 7 (and 28) days?	______	______

QUESTIONS

1. List two specific objectives of footing for a construction job.
2. In what type of location are retaining-wall footings often needed?
3. Discuss the earth-form method of footing.
4. Define the following terms.
   a. Footing.
   b. Forms.
   c. Specifications.
5. True or false: The first step in the pouring phase should be setting the steel rods within the form.
6. What is the difference between a standard mix and a dry mix? Which has better workability?

Chapter 3

Piers and Columns

Anchor bolts any of a variety of rather large J- or L-shaped bolts designed to have a portion embedded in concrete or mortar.
Anchor plates especially designed plate that has an anchor bolt welded to its undersurface.
Chamfer strips triangular corner strips used in forms to create a less sharp corner or edge surface.
Fluting a depression, round or otherwise, made in the length of a column.
Pier a support structure on which a horizontal beam or girder usually rests.
Scab a short piece of wood stock used as a brace.
Termite shield a specially formed metal shield that is placed over a pier or wall to prevent termite movement to wood stock.
Tie wires ordinary baling wire used to tie steel rods in the proper position.
Yoke a wood frame that surrounds a pier or column form (horizontally).
Yoke lock any of a variety (nails, braces, bolts) of devices used to hold a yoke together.

OBJECTIVES—INTRODUCTION

In Chapter 2 we saw that a pier's footing is separate from the wall and a pier must support a girder. A pier is synonymous with a column in that they both usually support some form of girder. If a difference exists, it is usually that a pier does not remain exposed to public view whereas a column does. In addition, a column may have architectural features such as fluting or chamfered corners.

The objectives for this chapter are two: *to be able to translate a pier* (or column) *blueprint into being, and to understand the methods of forming and the needs and uses of piers and columns.* The format of the chapter is the same as the format used in Chapter 2.

READING AND UNDERSTANDING THE PIER AND COLUMN BLUEPRINT AND SPECIFICATIONS

Three aspects of pier and column specifications are discussed. These are the foundation plan, for placement of the pier; the elevation plan, for design details; and specifications for metal and concrete requirements. Since the elements of piers and columns are nearly the same, the discussion centers around the pier, with only extraordinary column details added.

Foundation Plan and Placing the Pier

Recall that Figure 2–1 (p. 22) defined a need for piers through the center of the building and that the girder would be resting on top of the pier. The location of the center of the first pier in Figure 2–1 is 9 ft from the inside of the end wall (9 ft 8 in. from the outside) and 12 ft from the outside of the back wall. This location information is comprised of *specification data* taken from the foundation plan. Further examination of the foundation shows that a second pier is spaced 8 ft on center to the right of the first. Also, this plan identifies details CC and DD of the elevation plans.

Elevation Plan

As shown in Figure 3–1, detail CC illustrates the pier section of the full unit. Recall that Figures 2–4 and 2–5 provided footing data. In

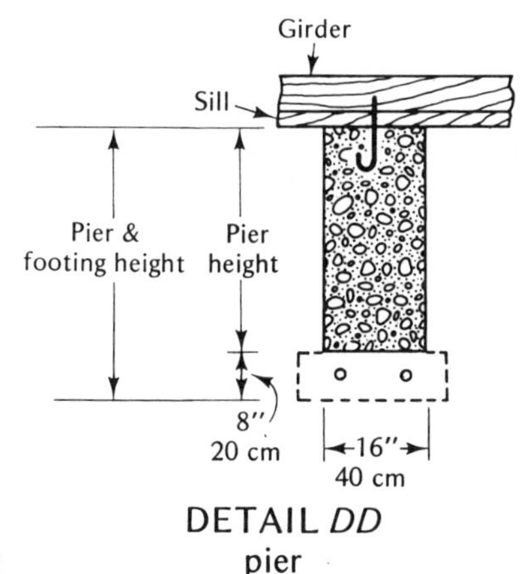

Figure 3–1 Pier, Elevation Drawing

Figure 3–1 the pier height and pier and footing height are included as specification details. In addition, the anchoring bolt position is included. Just how much of the bolt should be above the pier's top surface depends on the anchor bolt's function. Normally 3 to 3¼ in. of the bolt is exposed, to allow for:

a. Nominal-2-in. sill stock (part of girder).
b. Nut and washer.

Notice that the drawing does not show steel rods in the pier.

In short piers there is frequently no need for steel rods. The concrete has adequate compressive strength, and this strength adequately supports both live and dead weight loads, which are both projected downward. Therefore, there is very little need for added tensile strength.

In contrast, a high or tall pier or column does need steel reinforcement. Figure 3–2 shows an elevation detail plan of a high pier or column. In this drawing several specification factors are given. Two ½-in. steel rods are to be used to provide tensile strength. These two rods are to be tie-wired to two rods extending from the footing.

Also included in the details is an anchor bolt protruding from the

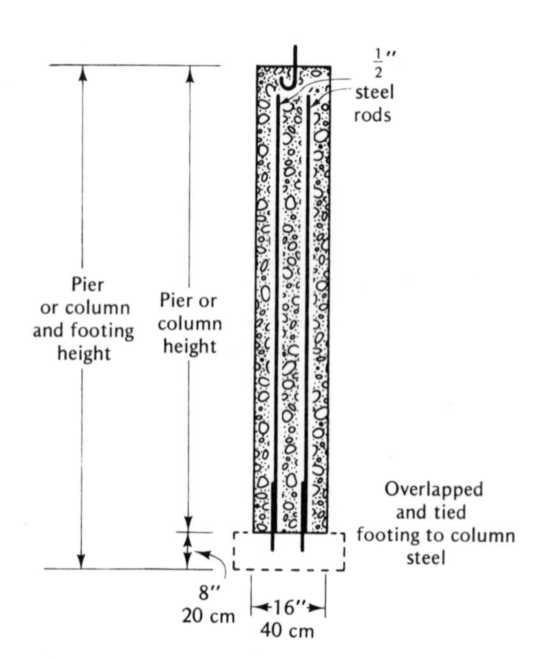

Figure 3–2 High Pier, Elevation Drawing

top of the column. Drawings probably will not include the anchor bolt, but the specifications will include a data line such as: "Anchor bolts or plates shall be ¾ in. in diameter (or 1 in. in diameter), 16 in. (or 18 in.) long and embedded all but 3 in., with on-center spacing of bolts."

Metal Products Needed for Piers and Columns

The anchor bolt has been mentioned several times. There are two distinct shapes usually used. Shown in Figure 3–3a is the L shape; in Figure 3–3b, the J shape. In addition, anchor plates with welded bolts may be needed, as Figure 3–3c and d shows. These are used where girder or truss assemblies are set into the plate and either nailed or lag-screwed to the plate.

In most residential and some commercial construction where piers are used, a termite shield is needed to protect wood products above the concrete level. A typical termite shield is shown in Figure 3–4. It rests on top of the pier, and its downsloping flanges prohibit termites from going past the shield. Also observe the use in this figure of an anchor plate.

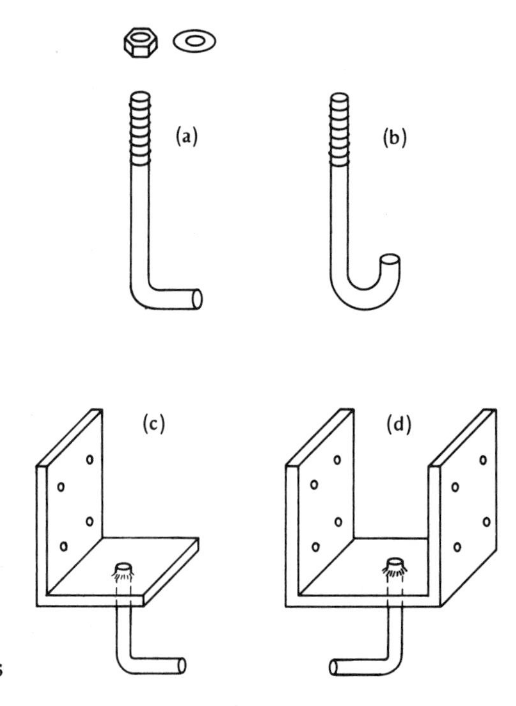

Figure 3–3 Types of Anchors

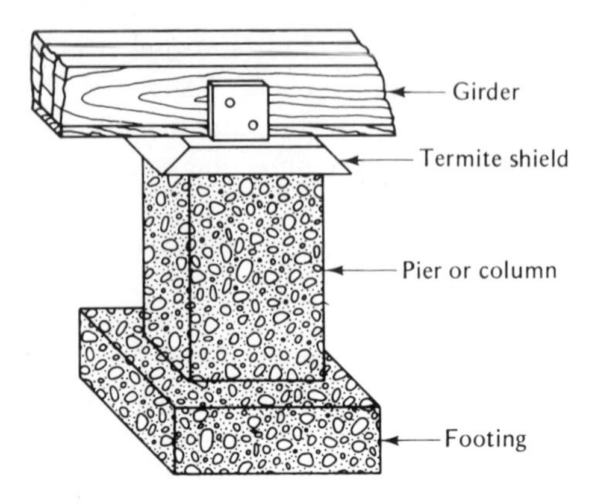

Figure 3–4 Termite Shield

Decorative Columns

Design details, as shown in Figure 3–5 or as written in specifications, are included where columns are exposed to public view. The modern trend is to use chamfer strips in the corners and triangle strips in flat areas to create flutes. Both are shown in Figure 3–5. The chamfered corners may or may not reach the full length of the column. Where the lower section (18 in. or so) of the column is to appear as a base,

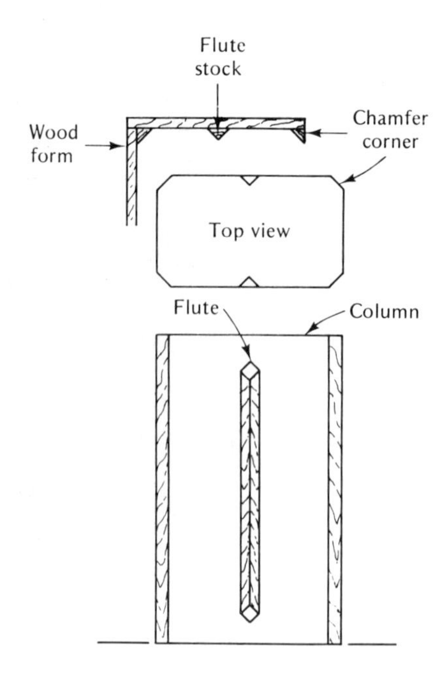

Figure 3–5 Flutes and Chamfered Corners

the chamfer strip does not extend to the floor. The same holds true if the top of the column is to appear to have a crown. In most cases the flutes do not extend to the floor or to the ceiling.

Quality of Concrete

In almost every application the concrete mixture will have aggregate size limited to a maximum of 1 in., although ¾ in. is more workable. The compressive strength must reach 2500 psi in 7 days and 3500 psi in 28 days. Compacting is required.

THEORY OF FORMING THE PIER OR COLUMN

Figure 3–6 contains all the details necessary for developing the theory of forming a pier and, for that matter, a column, except for the column's architectural character. The basic requirement that must be met by the

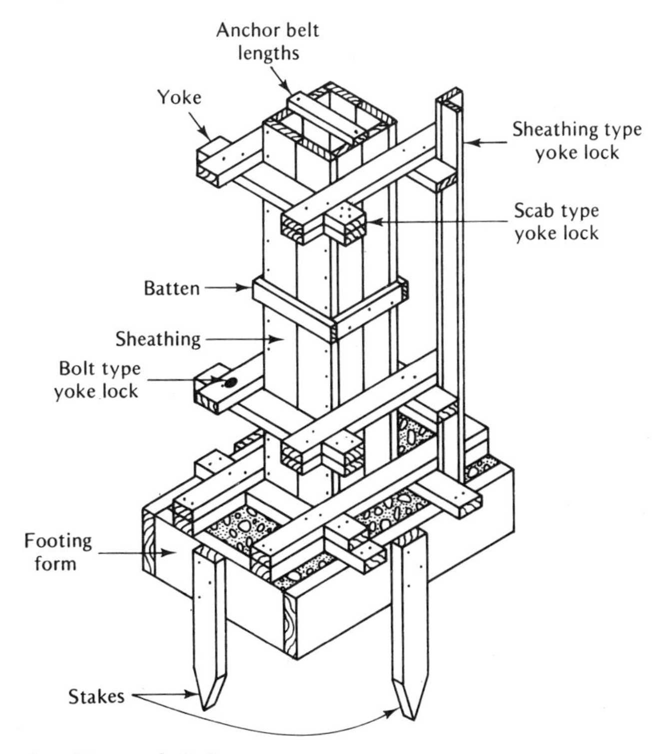

Figure 3–6 Form for Pier and Column

form is that the sides establish the outside parameters of the pier or column. The form may be constructed from nominal-1-in. stock or plyform (a form of plywood). The batten must be used if stock lumber is used.

Concrete's Pressure

Concrete in its wet and plastic states exerts a lateral (sideways) force on all four sides of the form; therefore, yokes usually constructed from nominal-2-in. stock are used to provide bracing. The concrete's force is distributed because the yoke is placed around the form.

Yokes

Yokes are placed, one at the bottom, one near the top, and intermediate ones at intervals depending upon the height of the form, the thickness of

TABLE 3–1 CONCRETE PRESSURES FOR COLUMNS AND WALL FORMS (COURTESY OF AMERICAN PLYWOOD ASSOCIATION)

Pour Rate (ft./hr.)	*Pressures of Vibrated Concrete (psf) (a), (b), (c)*			
	50° F		*70° F*	
	Columns	*Walls*	*Columns*	*Walls*
1	330	330	280	280
2	510	510	410	410
3	690	690	540	540
4	870	870	660	660
5	1050	1050	790	790
6	1230	1230	920	920
7	1410	1410	1050	1050
8	1590	1470	1180	1090
9	1770	1520	1310	1130
10	1950	1580	1440	1170

Notes: (a) Maximum pressure need not exceed 150h, where h is maximum height of pour.
(b) For non-vibrated concrete, pressures may be reduced 10%.
(c) Based on concrete with density of 150 pcf and 4 in. slump.

TABLE 3–2 YOKE SEPARATION ON PIER OR COLUMN FORM

Largest Dimension of Column in Inches = 'L'

Height: 1′, 2′, 3′, 4′, 5′, 6′, 7′, 8′, 9′, 10′, 11′, 12′, 13′, 14′, 15′, 16′, 17′, 18′, 19′, 20′

L	Yoke separations (from top down)
16″	31″, 31″, 30″, 29″, 21″, 20″, 18″, 15″, 14″, 13″, 13″, 12″
18″	29″, 28″, 28″, 26″, 20″, 18″, 16″, 15″, 13″, 12″, 12″, 11″, 11″
20″	27″, 26″, 26″, 24″, 19″, 16″, 15″, 14″, 12″, 11″, 11″, 10″, 10″, 10″, 9″
24″	23″, 23″, 23″, 22″, 16″, 14″, 13″, 12″, 11″, 10″, 9″, 9″, 9″, 9″, 8″, 8″, 8″
28″	21″, 21″, 20″, 18″, 15″, 13″, 12″, 10″, 9″, 9″, 9″, 8″, 8″, 8″, 7″, 6″
30″	20″, 20″, 19″, 18″, 18″, 12″, 12″, 10″, 9″, 8″, 8″, 8″, 7″, 7″, 7″, 6″
32″	19″, 19″, 18″, 17″, 13″, 12″, 10″, 10″, 8″, 8″, 8″, 7″, 7″, 7″, 6″, 6″
36″	17″, 17″, 17″, 15″, 12″, 11″, 10″, 8″, 8″, 7″, 7″, 7″, 6″, 6″, 6″

L

the form walls, and the pour rate of concrete. For example, if the pour rate is established at 2 ft/hr, the pressure on the form would be 410 lb/ft^2 (pounds per square foot or psf), as shown in Table 3–1. Increasing the pour rate to 8 ft/hr, for instance, increases the pressures on the form to 1180 lb/ft^2. This type of pressure creates a need for additional yokes on the form, additional bracing, and greatly increases the chances for the form to lift from its base and collapse. The best alternative is to establish a reasonable pour rate and construct the form accordingly.

Yoke separation should follow the schedule shown in Table 3–2. The distances between yokes are found by (1) locating the form height in the left column, (2) locating the form length across the top, and (3) intersecting these two points in the table (e.g., pier 8 ft in height with length = 16 in., yoke spacing is 30 in. on-center).

Pier and Column Forming Alternatives

The first of two methods of forming a pier or column form is to pour the footing first, then construct the pier form and install it on top of the footing. If the footing forms are left intact, they may be used as the nailing surface for the base yoke. Diagonal bracing for plumb may also be nailed to the footing form, although to be effective on tall piers or columns, the bracing must extend out farther.

In the second method the footing and pier or column are poured in one operation. Both forms are constructed at the same time, set in position, staked, and braced. Then the concrete is poured in place. In many operations of this kind the top surface of the footing is completely covered with form material so that the concrete does not bulge out above the footing.

Architectural Additions to the Forms

Any number of characteristics may be added to the inside of a form to create architectural character. A few are the chamfer strip, half-round stock, triangle stock, flat squares with beveled sides, and circles with beveled edges (Figure 3–7). Notice that all pieces are made for easy removal and that the depression created is one that sheds water easily in the event the column is exposed to the weather.

All these pieces of stock are nailed securely to the form so that when the form is removed, they come away from the concrete with the form.

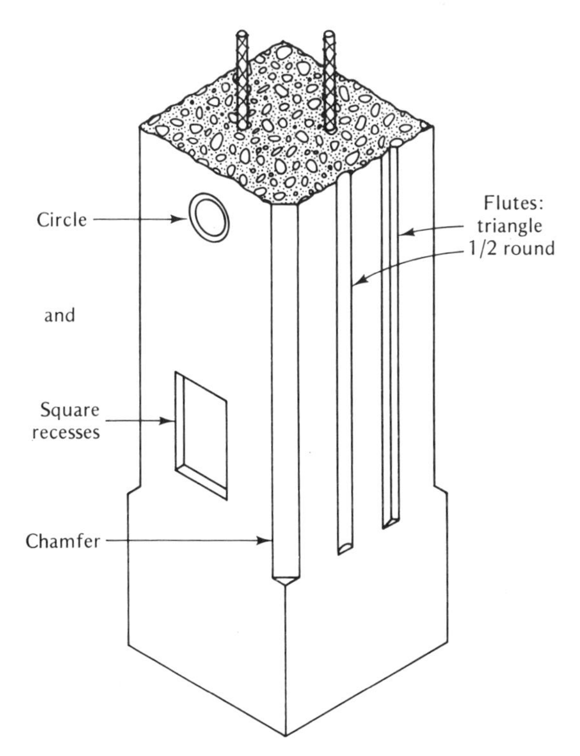

Figure 3–7 Architectural Refinements for Columns and Piers

THEORY OF POURING AND FINISHING PIERS AND COLUMNS

Several steps are usually associated with pouring the concrete. One is performed prior to building the form—setting the reinforcement rods. Another is pouring and compacting the concrete, and a third is setting anchor bolts or plates.

Setting Reinforcement Rods

Figure 3–8 illustrates a precast footing with two steel rods embedded. If this technique is used, no steel would be set, but location or centering of the rods within the form would be ensured.

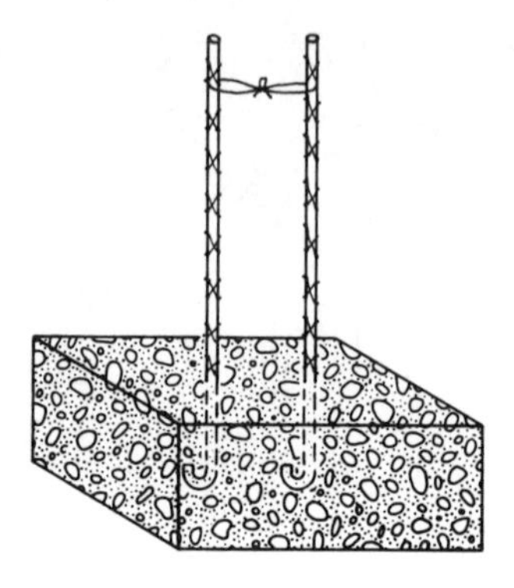

Figure 3–8 Rods in Footing

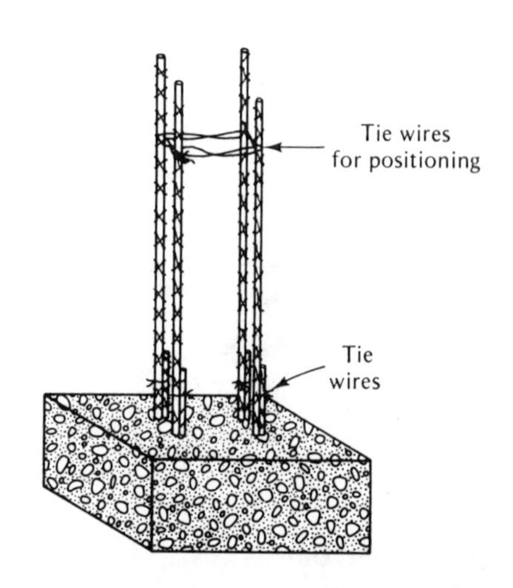

Figue 3–9 Rods Added After Footing Is Poured

In contrast, Figure 3–9 shows that the rods must be cut, positioned, and tied prior to erection of the form. Four rods are shown; however, the same technique can be used for two-rod reinforcement.

Pouring and Compacting the Concrete

Recall that the form is built to sustain a specific pour rate. The rate should be maintained to preclude structural collapse of the form as well as lifting of the form from its base. As the concrete is lowered into the form, a mechanical vibrator may be used as a compactor; or, on small piers, for example, a short length of 2 × 4 stock may be used to tamp the concrete into the form.

It is especially important that some form of compacting be done to guarantee that concrete fills all the crevices and encloses all the surfaces of the reinforcing rods. However, excessive vibration causes separation of the concrete's cement aggregate and water, so use caution. Finally, the concrete at the top of the form should be trowelled smooth.

Installing Anchor Bolts

Refer again to Figure 3–6 and notice that there is a short piece of lumber stock nailed across the top of the form, labeled *anchor bolt template*. This piece of stock is predrilled to allow passage of the bolt through its thickness. The washer and nut are installed so that they resemble those in Figure 3–10. As many bolts as are needed may be so located and installed.

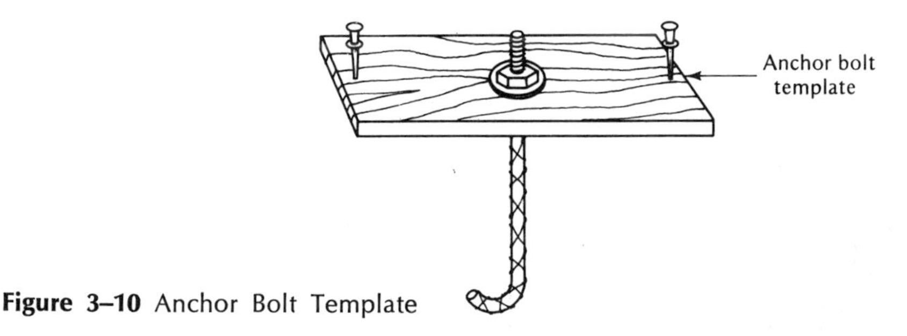

Figure 3–10 Anchor Bolt Template

When this piece of stock is ready with bolt or bolts inserted, the unit is positioned on top of the pier and the bolts are worked into the freshly poured concrete. Two nails hold the template in position.

Removing Forms and Finishing Piers or Columns If Required

Forms that have been built to be reused must be removed carefully so that they are not damaged. Normally, the process used is just the reverse of the construction process:

- **a.** Remove the braces.
- **b.** Remove the yokes.
- **c.** Remove the sides of the form.
- **d.** Remove the footing forms and stakes.

While these forms and other pieces are being handled, they should be cleaned of nails and any residue of concrete. The inside surfaces should also be oiled at this time. Then all materials should be moved to the next location or stored safely.

Grouting or plastering may be needed on all surfaces of the columns and piers that will remain visible. Mixtures may include white sand and white cement or mortar mixture and fine sand. The contract specifications will usually stipulate the desired finish. It is probable that the specifications will also dictate the type of surface finish, such as floated, brushed, or smooth.

Work Flow Summary

This illustrated analogy of the masonry work flow shows that the masons are frequently interrupted by the carpenters' building forms, because masons rarely build their own forms. It is also possible that steelworkers may set the reinforcement rods. However, pouring, compacting, and finishing the pier or column is exclusively masons' work.

INSPECTION OF THE PIER OR COLUMN

Inspection	*Satisfactory/N.A.*	*Unsatisfactory*
Has the pier or column been positioned according to the foundation plan?	______	______
Has the form been built with proper:		
a. Width?	______	______
b. Height?	______	______
c. Thickness?	______	______
d. Yoke separation?	______	______
e. Adequate bracing?	______	______
f. Plumb to ±¼ in.?	______	______
Have the reinforcement rods been installed, positioned, and tied?	______	______
Has the proper concrete mixture been contracted for, delivered, and used?	______	______

Inspection	*Satisfactory/N.A.*	*Unsatisfactory*
Has compacting been performed properly?	______	______
Have all the anchor bolts been installed properly?	______	______
Have proper curing techniques been used?	______	______
Does the final finish meet the qualitative and architectural requirements of the contract?	______	______

QUESTIONS

1. What is the difference between a pier and a column?
2. List the two anchor bolt shapes and how they are commonly used.
3. What is the maximum aggregate size of a concrete mixture? What size is the most workable?
4. Define a yoke and describe its purpose.
5. How are the distances between yokes determined?
6. List several characteristics that may be used to give architectural character to a form.
7. What three steps are involved in pouring concrete?
8. True or false: Pouring, compacting, and finishing the pier or column is exclusively carpenters' work.

Chapter 4

Pilasters

Bracing nominal-1-in. and 2-in. stock used in form building to steady and hold the forms in proper position.

Buckle in forms, structural breakdown of any portion of the form or its bracing caused by the pressure of the concrete.

Encapsulate in pouring concrete, the requirement for total contact (surrounding) of concrete with each metal reinforcing bar.

Flush putting two surfaces in line.

Forming the task of building forms that will contain concrete in specific dimensions.

Plumb a perpendicular alignment usually made with a level or plumb bob and a line. A requirement for all form work.

Stud a nominal-2-in. × 4 in. member of wood in any length.

Ties wire, rods, or other materials used in form making to establish and maintain separation of form walls.

Wales nominal-2-in. lumber stock positioned horizontally at specific vertical intervals along a wall form.

OBJECTIVE—INTRODUCTION

A wall built of its own material, concrete or other, that is of considerable length generally has an unsound character. Wind and other natural forces cause it to move, swell, shrink, and eventually to buckle or crack. A *pilaster* inserted at periodic distances provides the wall with rigidity, stability, and added strength.

A second and equally important function for the pilaster is its use as a pier. In this application it is designed and located to provide support and anchoring for a girder, header, or truss roof assembly. Figure 4–1 shows several varieties of application. Based on these pur-

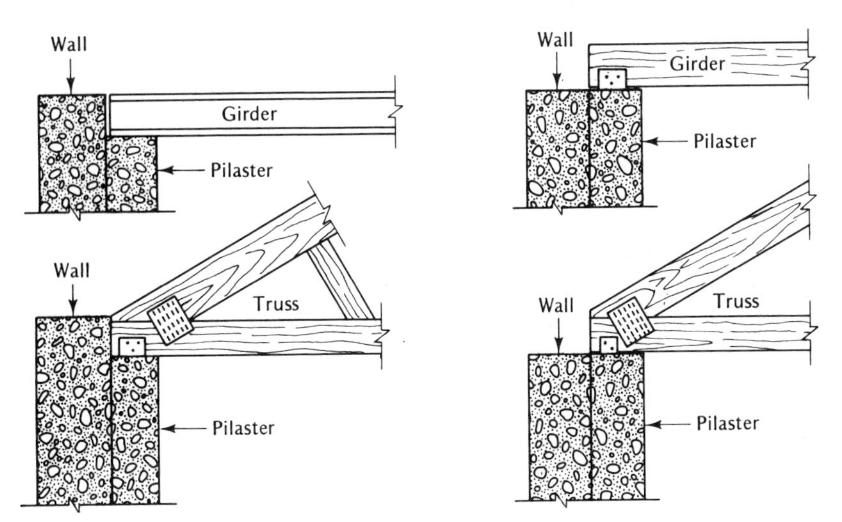

Figure 4–1 Pilaster Used as a Pier

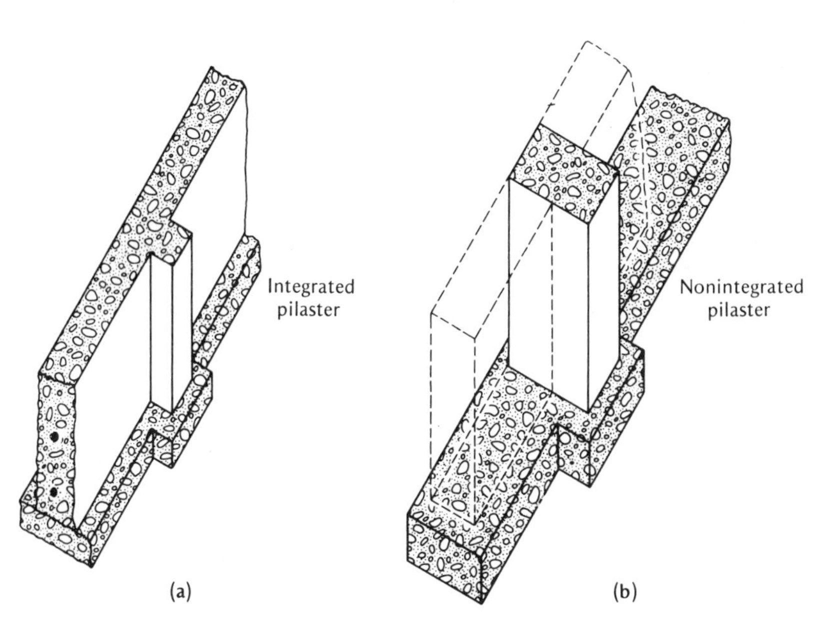

Figure 4–2 Pilasters (Pictorial)

poses, one objective covers all necessary aspects: *to be able to translate a blueprint and specifications of a pilaster into being.*

Modern building techniques vary greatly, so two methods of building a pilaster are included. One is labeled the *integrated pilaster,* as shown in Figure 4–2a. Notice that it projects inward from the wall as all pilasters do, but, more important, that all concrete for wall and pilaster is poured at the same time.

In contrast, the other method studied here is the *nonintegrated* or *stand-alone pilaster.* Its method of construction, which is much like that of a pier, is shown in Figure 4–2b. The dashed areas show that the walls are butted against the pilaster.

The discussion to follow covers specification and blueprint data, forming and pouring theory, and an inspection checklist prepared for practical use.

READING AND UNDERSTANDING A BLUEPRINT AND SPECIFICATION DATA

Again, as in the two previous chapters, three elements—placement, dimensions, and specification data not included in plans—are examined.

Foundation Pilaster Data

Refer again to Figure 2–1 (p. 22), which shows the partial foundation plan, and locate detail BB. Several factors (specifications) obtainable from this plan relate directly to the pilaster:

1. Its location is 12 ft (3.6 m) on-center from the outside of the back wall line.
2. It projects 8 in. (20 cm) into the building from the wall line.
3. Its footing extends on three sides and is 16 in. (40 cm) wide and 12 in. (30 cm) inward. (The footing data can be found in Chapter 2.)
4. Detail BB directs attention to the elevation plan discussed later.

The plan shows that this pilaster is to be an integrated type, which means that it is poured along with pouring the wall. The complexity of its construction is illustrated in the section on forming theory.

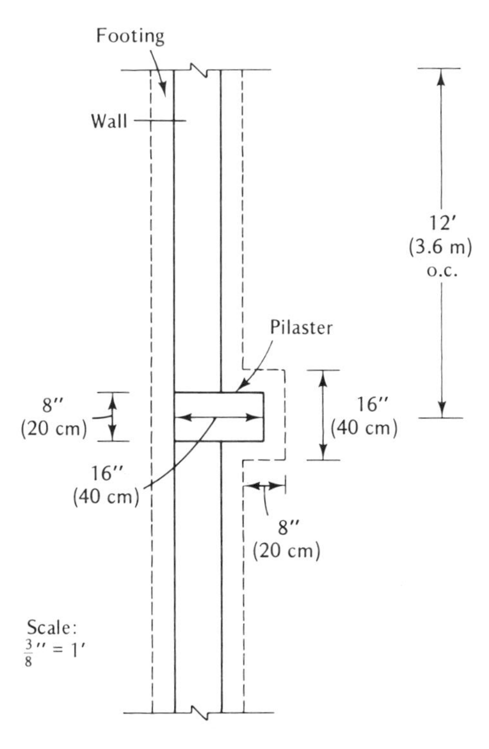

Figure 4–3 Foundation Plan—Stand-Alone Type Pilaster

If a stand-alone, nonintegrated type were required, its data could be extracted from a similar foundation plan section or from a foundation plan similar to the section shown in Figure 4–3. In the section of a foundation plan shown in Figure 4–3, the pilaster data consist of its location to the back wall, its flush surface with the outside of the wall that it will form a part of, its dimensions, and the dimensions of its footing. The main difference in this drawing is the implication that it must be constructed prior to constructing the wall sections on either side of it.

Elevation Plan Pilaster Data

Detail BB in the foundation plan identifies the pilaster elevation plan notation. Figure 2–3 (p. 24) illustrates the elevation footing requirements for the pilaster, and Figure 4–4 details the pilaster dimensions.

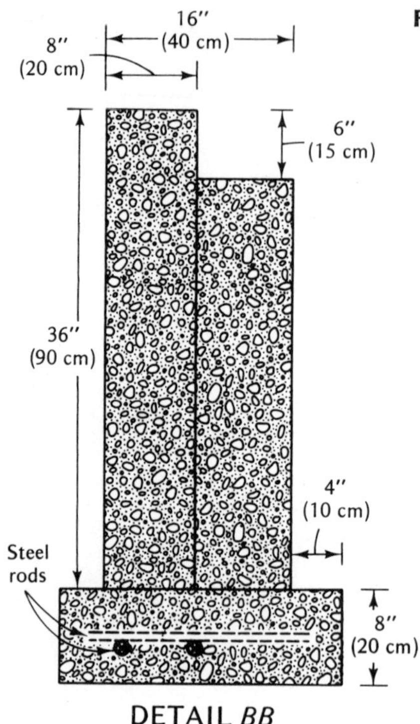

Figure 4–4 Pilaster—Wall Dimensions

A careful study of the plan detail shows:

1. The wall height is 36 in. (90 cm) above the footing.
2. The wall thickness is 8 in. (20 cm).
3. The pilaster portion is only 30 in. (75 cm) high, thereby creating a 6-in. or 15-cm offset depression for placement of the girder.
4. The overall width of the pilaster is not shown [the foundation plan gave this dimension as 8 in. (20 cm)].
5. The overall wall pilaster length is 16 in. (40 cm).
6. The pilaster footing extends 4 in. (10 cm) beyond the pilaster's inner surface.

In this plan the pilaster serves two purposes: it strengthens the 24-ft-long wall, and it provides support for the girder that is needed through the center of the building.

Refer again to Figure 4–1 (p. 51) and note the two typical arrangements of pilaster height. The two detail plans on the left show an offset depression. The one indicating a girder has just been detailed, but notice that this arrangement is also used where a roof truss assembly is set and secured to an anchor plate. The other two examples show pilasters with heights equal to wall height, and the girder or truss rest on the top. Usually these resting components do not extend into the wall, but there may be occasions where the plans require it.

Specification Data

Data not included in drawings are listed in specifications. For the crawl-space plan and pilaster requirements, a data line, "pour pilaster along with wall to the same concrete specifications and include reinforcement bar (or bars) within the pilaster area," would probably be included by the architect if the foundation is located on shifting soil or in an earthquake-prone area.

However, there may not be a data line for this job because the height of the pilaster is only 30 in. (75 cm) and the concrete would have sufficient compressive strength to support the weight of the girder, floor joists, and live and dead loads. Another reason why no data could be listed in the specifications is that both the carpenter

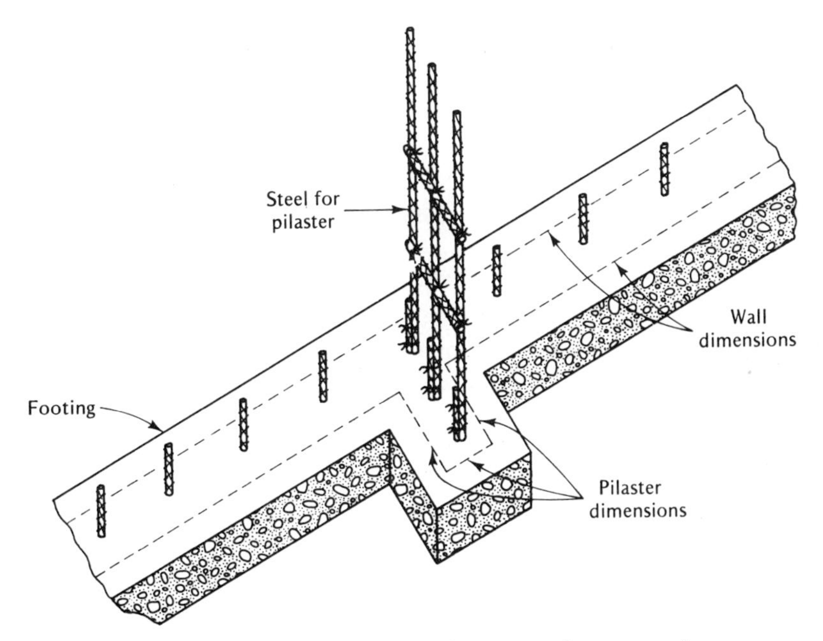

Figure 4–5 Stand-Alone Pilaster Steel Bar Needs

who builds the forms and the mason who pours and finishes the wall know what needs to be done from reading the blueprint.

Specifications for the stand-alone pilaster, especially one of full room height, 8 ft or more, would include steel-rod requirements. Figure 4–5 illustrates what the specifications could require. Bars of reinforcing steel could be projecting from the footing both along the wall and from the pilaster section, or just for the pilaster. These data would be a specification. The number and size (dimension) of bar would also be given.

THEORY OF FORMING A STAND-ALONE PILASTER

The method of building a stand-alone or nonintegrated pilaster is the same as building a pier. The physical dimensions of height, width, and length are taken from the elevation and foundation plans and are used by carpenters to build the sides of the form and the yokes. The pieces of form are assembled, positioned, and secured with bracing so that the liquid concrete does not move the form in any manner.

Just as in forming piers, forming for a pilaster must be built for a specific pour rate, and both carpenter and mason must know the rate. Recall that yoke spacing was based upon pour rate and that fast pouring rates create great pressure. Ensuring stability and plumb accuracy of the form must be a part of the design both in building and in withstanding the pressure of liquid concrete.

THEORY OF FORMING THE PILASTER AS PART OF A POURED WALL

Figure 4–6 shows the complex building requirements used by carpenters to build an integrated pilaster form. The one shown is constructed of plyform material and is secured in position by using various techniques. Locate the following items in Figure 4–6:

1. Pilaster tie.
2. Pilaster wales.
3. Pilaster form.
4. Pilaster footing.
5. Wall form.
6. Studs.

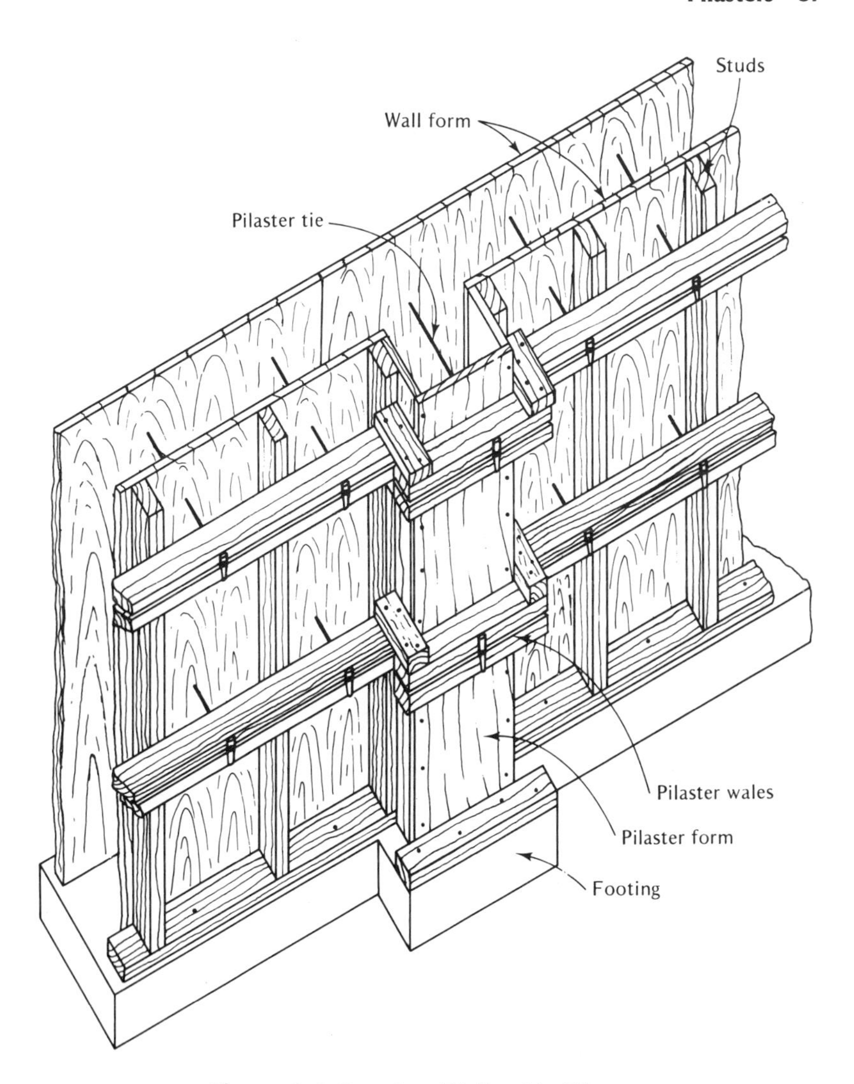

Figure 4–6 Forming Wall with Pilaster

The pilaster tie is similar to a wall tie except that it is longer. There are various styles of ties, including wire, but the one shown is a snap tie and is illustrated in Figure 4–7. It consists of the tie rod washer and a clamp.

Pilaster wales are short lengths of 2×4s that act as yokes. Notice that the tie passes between the wales and the clamp secures the wales to the pilaster form.

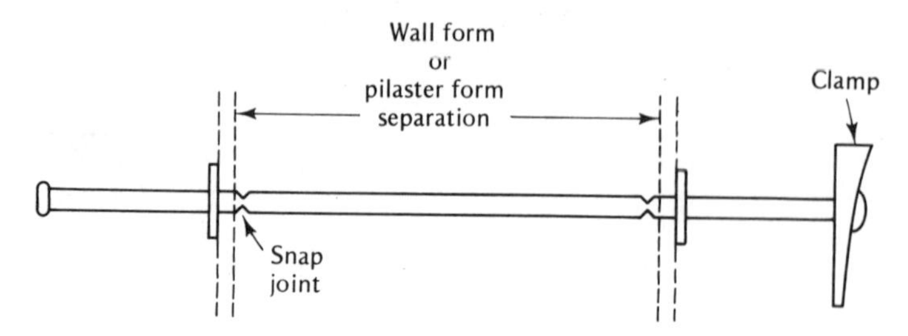

Figure 4–7 Snap Tie

The pilaster form is made from three pieces of plyform plywood, two side pieces and one face piece. These are nailed together with either 8d common nails or double-headed nails especially made for form work.

Also note the short length of the 2 × 4 nailed to the footing. It is important that this piece be installed. However, should the footing forms still be in place, this piece could be nailed to them rather than to the concrete footing.

During construction of this form, the critical factors are proper inside dimensions for width, length, and height; secure nailing of form pieces; and proper bracing. The pour rate shall be a controlling factor in the number of wales used and their on-center separation. Also of vital concern is the requirement for vertical accuracy (plumb). This must be established, and bracing (not shown in Figure 4–6) must be included prior to pouring the concrete.

THEORY OF POURING AND FINISHING PILASTER

Three elements are included in the theory of pouring concrete: setting steel if used, pouring the concrete correctly, and finishing the pilaster once the forms are removed.

Setting Steel

When using the stand-alone pilaster as in the example shown in Figure 4–2b, steel reinforcing bars are usually needed. They may be tied to short lengths of bars protruding from the footing. However, this is not always the case. The bars may be free-standing within the area of the pilaster.

In any case, they should be positioned equidistant from form sides so that sufficient concrete thickness is obtained between bar and form. They should also have uniform separation so that concrete may encapsulate their entire surface.

In an integrated pilaster, one or two reinforcing bars may be installed within the pilaster, or there may be a continuous web through the length of the wall. Where bars are used, they are joined to the wall steel by horizontal cross-bars and/or tie wires.

Pouring Concrete

Recall that the sample plan used for explanation was a crawl-space foundation wall only 3½ ft high. Pouring concrete into this form means that the concrete falls a maximum of 3½ to 4 ft, so little separation should occur. However, in full-height walls and stand-alone pilasters, the concrete may drop 8 ft or more. This situation may cause separation of the mixture. Avoid this condition at all times by lowering the concrete into the form.

Coincident with filling the form, there must be manual or mechanical *compacting*. Manually, lengths of 2 × 4 or 2 × 3 stock are used to tamp the concrete at, for example, 2-ft intervals during the pouring. Mechanical means, such as the use of an electric vibrator, are quicker and very efficient but must be done cautiously, as strong agitation may separate the concrete mixture.

Above all, do not forget that the form was built to sustain a specific amount of lateral pressure. This restricts pouring to a definite pour rate. Pouring faster than planned for places added forces against the form and could cause the form to rupture, to shift out of alignment or out of plumb (tilting), or could lift the form from its seat. Any of these problems will result in a structural defect.

Finally, check the plans and specifications for data on anchor devices that may be needed, such as girder or truss plates, anchor bolts, or even reinforcing bars needed to join pilaster to poured concrete headers or girders.

Removing the Forms and Finishing the Pilaster

Removing the form is a process opposite to constructing it. As the yoke or wales are removed, they should be set aside and cleaned for possible reuse. Form pieces should be cleaned of concrete residue and oiled.

Tie rods need to be snapped off below the surface of the concrete at the place designed for the break. Where the tie is snapped off, a small hole remains and it, as well as any other surface holes or gaps in the concrete, needs to be plastered. If a special textured surface is needed, it, also, is done at this time so that the pilaster is fully completed and workmen need not return to it later for further work.

INSPECTION OF THE PILASTER

Inspection	*Satisfactory/N.A.*	*Unsatisfactory*
Has the pilaster been properly located from drawing details?	________	________
Has the form been erected to the proper dimensions and in position?	________	________
Is it plumb and properly braced?	________	________
Has the proper type of concrete been delivered and placed into the form?	________	________
Has compacting been done?	________	________
Have curing procedures been followed?	________	________
After form removal, has the pilaster been finished according to specifications?	________	________

QUESTIONS

1. Define the following terms.
   a. Encapsulate.
   b. Plumb.
   c. Ties.
   d. Wales.
2. What is the function of a pilaster?
3. How may a pilaster differ from a pier?
4. Explain the difference between a stand-alone and an integrated pilaster.

5. What is the need for steel reinforcing rods in a pilaster?
6. Explain the need for maintaining a pour rate and the results that can occur if the pour rate is exceeded.
7. Should rods be installed against the form of a pilaster? Why not?
8. How could you compact concrete placed into a pilaster form?
9. How could exceeding the pour rate affect the pilaster form?
10. What should be done with the holes left after the ties are removed?

Chapter 5

Concrete Walls

Battens nominal-1-in. stock used to aid in holding other members and braces.
Casting the act of depositing concrete as though throwing it.
Deflection a bend caused by stress or pressure at a point.
Grouting a material used to fill holes in concrete, or the act of installing or applying the grouting mixture.
Joint any place where two or more edges or surfaces come to a union (e.g., plywood to plywood, concrete to concrete).
Lateral force side-to-side force.
Offsets depression-constructed plywood with surface materials added.
Plumb vertically perpendicular, as measured with a spirit level or plumb bob.
Sheathing nominal-1-in. stock, plywood or plyform, used in forming.
Shoe plates nominal-2-in. stock placed on top of or near the footing; part of a form near its base.
Strong back a heavy member placed perpendicularly against the wales; a part of the form's bracing.
Studs nominal-2-in. stock used to support the form-wall materials.
Ties wire, rod or snap, used to hold wall forms at specific separation.
Wales horizontal members that aid in wall-form reinforcement and distribution of forces.

OBJECTIVES—INTRODUCTION

For the mason, probably the most important aspect of building a reinforced concrete wall is the adequacy and accuracy of the form. Where the form is soundly constructed with studs, wales, and ties spaced properly, using the proper thickness of sheathing or plywood, the mason can fill the form and be reasonably sure of the finished product. If perfect perpendicular and horizontal positioning and wall thickness are maintained with the aid of bracing, even greater assurance of the accuracy of the finished product is possible.

Other characteristics of the wall, such as openings for windows, vents, piping, and the like, are added during forming. When the concrete is poured into these forms, it must encapsulate the openings and steel reinforcement as well as fill the entire form. These important factors lead to the objectives of this chapter: *to understand the requirements for accurate, sound wall forms;* and *to achieve a knowledge of the properties of concrete when pouring a wall and during curing.*

THEORY OF FORMING A WALL

Several types of wood forms are examined in this section, since each is constructed differently. General requirements for these forms are explained to show how they perform their function. Then the usual assembly methods are given so that knowledge in this area is obtained for future use. But first blueprint and specification data are examined to illustrate how very few data are provided to the carpenter building the form or to the mason pouring the form.

Blueprint and Specification Data

Figure 5–1 shows a typical foundation blueprint for a residential structure. Figure 5–2 shows an elevation detail plan for a wall in this house. Together, all dimensions of all foundation walls are given:

1. From Figure 5–1, the foundation plan:
   a. Position of walls.
   b. Length of each wall.
   c. Wall thickness.
2. From Figure 5–2, elevation detail data:
   a. Wall height (note that the example shown states "varies"; in each specific plan this entry would include a dimension placing the footing well below the frost level).
   b. Offsets in the wall for brick facing.
   c. Anchor bolt detail.

Significantly absent are data on form-building or reinforcement requirements. In some cases, especially where seismic conditions must

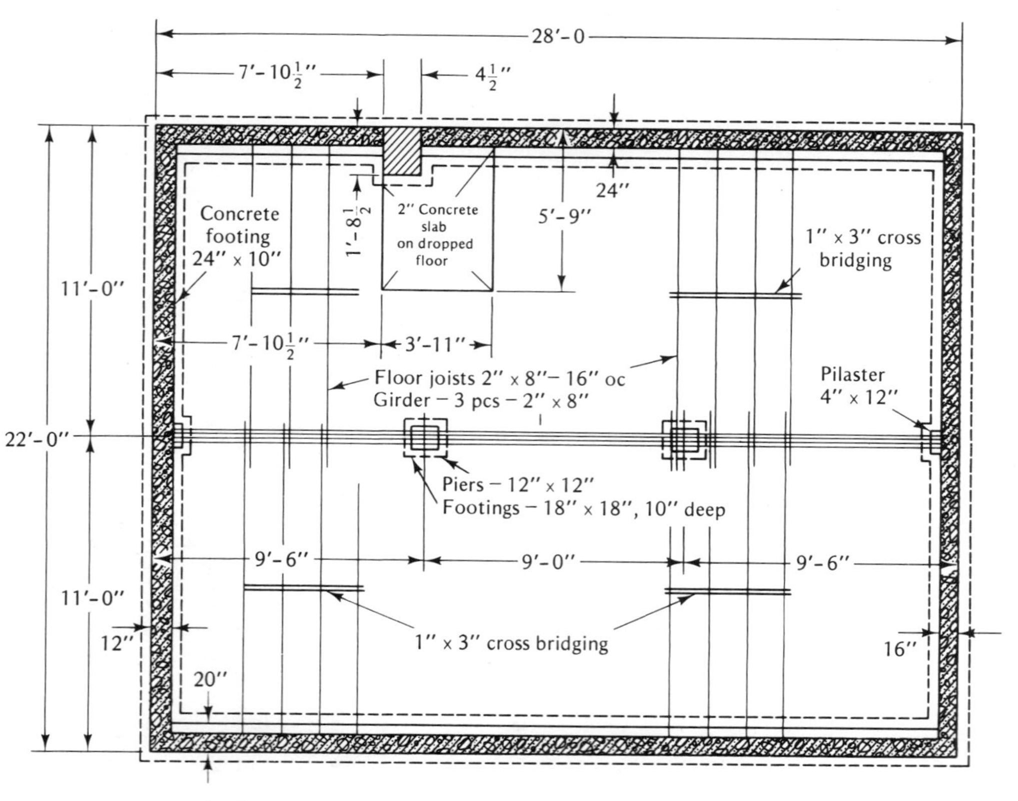

Figure 5–1 Sample Foundation Blueprint

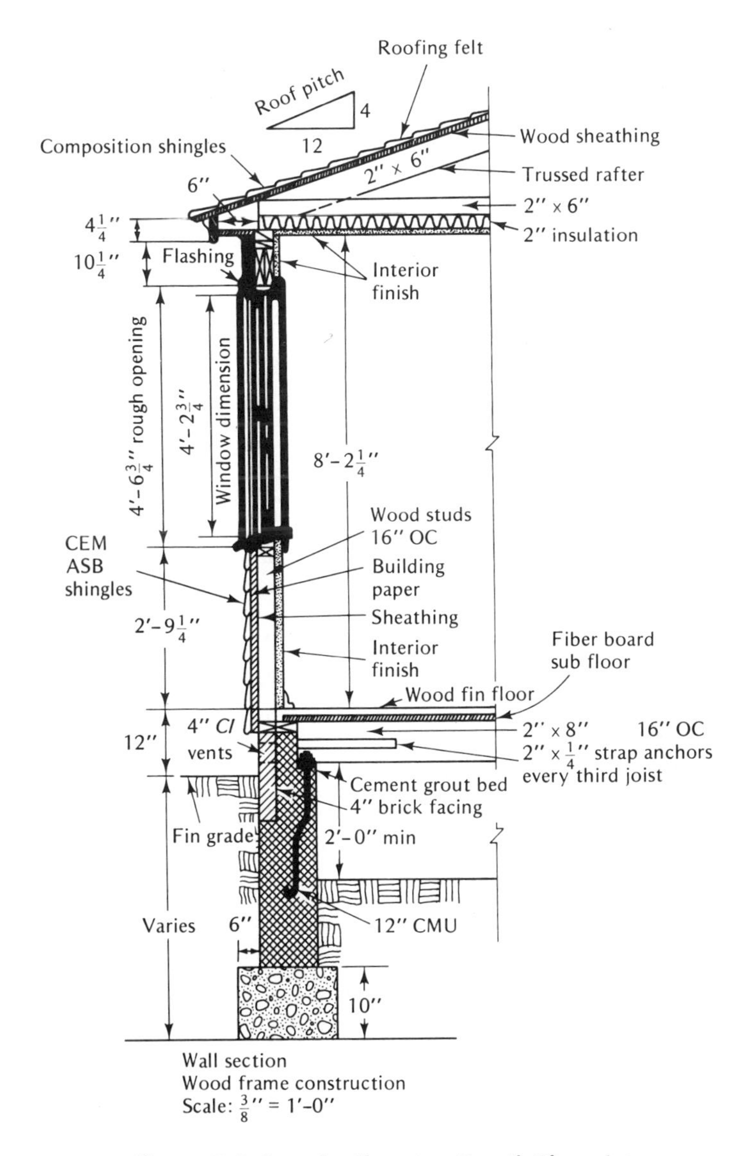

Roofing felt
Roof pitch
4
12
Composition shingles
Wood sheathing
2" × 6"
Trussed rafter
6"
2" x 6"
4¼"
2" insulation
10¼"
Flashing
Interior finish
4'-6¾" rough opening
4'-2¾"
Window dimension
8'-2¼"
Wood studs 16" OC
CEM ASB shingles
Building paper
Sheathing
2'-9¼"
Interior finish
Fiber board sub floor
Wood fin floor
12"
4" CI vents
2" x 8" 16" OC
2" x ¼" strap anchors every third joist
Cement grout bed
4" brick facing
Fin grade
2'-0" min
Varies
6"
12" CMU
10"
Wall section
Wood frame construction
Scale: ⅜" = 1'-0"

Figure 5–2 Sample Elevation-Detail Blueprint

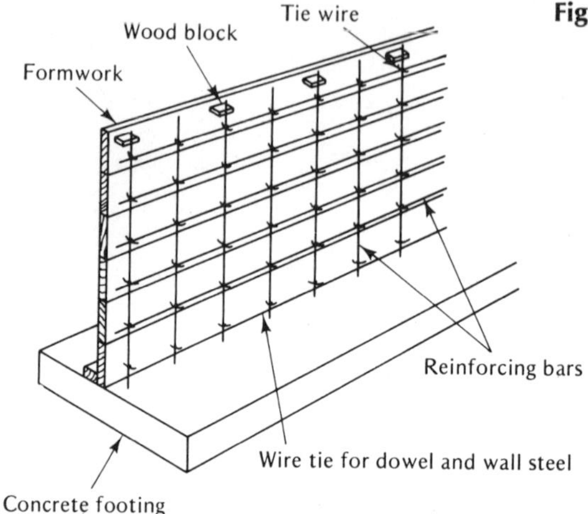

Figure 5–3 Steel Placement

be considered, a steel placement schedule will be drawn up (Figure 5–3). But more probably a specification data item such as one of the following is given:

SPEC. ITEM: Wall reinforcement: use bar No. 4 (½ in. diam) spaced horizontally each 16 in. o.c. and vertically each 12 in. o.c.

SPEC. ITEM: Wall reinforcement: use bar No. 6 (¾ in. diam) placed horizontally and vertically 24 in. o.c.

SPEC. ITEM: Wall reinforcement: use welded wire fabric grade _______, size 6 in. × 6 in. (Figure 5–4).

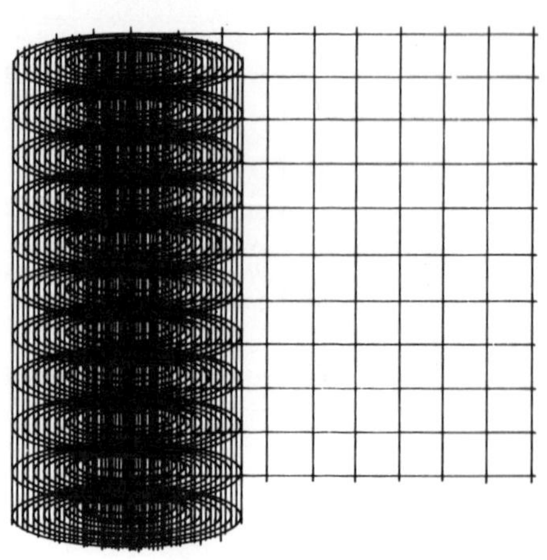

Figure 5–4 Welded Wire Fabric

The data extractable from the plans and specifications may seem incomplete to the layman and apprentice workman, but it is usually sufficient for a journeyman. He knows through on-the-job training and courses he has taken that the concrete used in wall construction usually weighs from 110 to 180 lb/ft^3. So dead weight and lateral force of concrete are considerable. The carpenter building the form must know the capability of his materials to sustain the lateral force of the concrete while it is in its plastic state, as well as its dead weight. By addition of these factors, planning and construction of the form is possible.

General Requirements for a Wall Form

The following nine general requirements should make clear that the integration of sound building practices and quality materials produces a sound wall form. When all of these are present and can be adequately evaluated for their quality, accuracy, and adequacy, the mason knows the requirements during the pouring phase of building the concrete wall.

1. *Types of materials:* Structurally sound members, fir and pine grade No. 2, should be used for wales, studs, braces, shoe plates, strong backs, mudsills, and stakes. The same grade of nominal-1-in. sheathing should be used for form walls. Where plywood or plyform [1] is used, its grade should be coincident with other form needs for adequacy. Ties, spreaders, and the like must meet the tensile-strength needs of the form—that tensile strength exceed the force of plastic concrete. Ties, nails, and bolts should have a stated minimum holding ability.

2. *Location and size of members:* The four examples provided in Figure 5–5 illustrate the uses of the materials. Note in Figure 5–5a that double wales are used and that the wall tie passes between them and is held fast by a clamp. This technique was explained in Chapter 4. The figure also shows a *shoe plate* on both sides of the wall. Frequently, only one side has a shoe plate. The ties, wales, and studs provide adequate support for the plyform or sheathing materials.

In Figure 5–5b, wire ties are used, and sheathing is held at proper separation by wood spreaders. The shoe, wale, and stud are used, and a strong back forms part of the bracing system.

In Figure 5–5c studs are braced individually, and sheathing or plywood nailed directly to studs maintains separation with the aid of the tie across the top of each stud.

---

[1] *Plywood for Concrete Forming,* American Plywood Association.

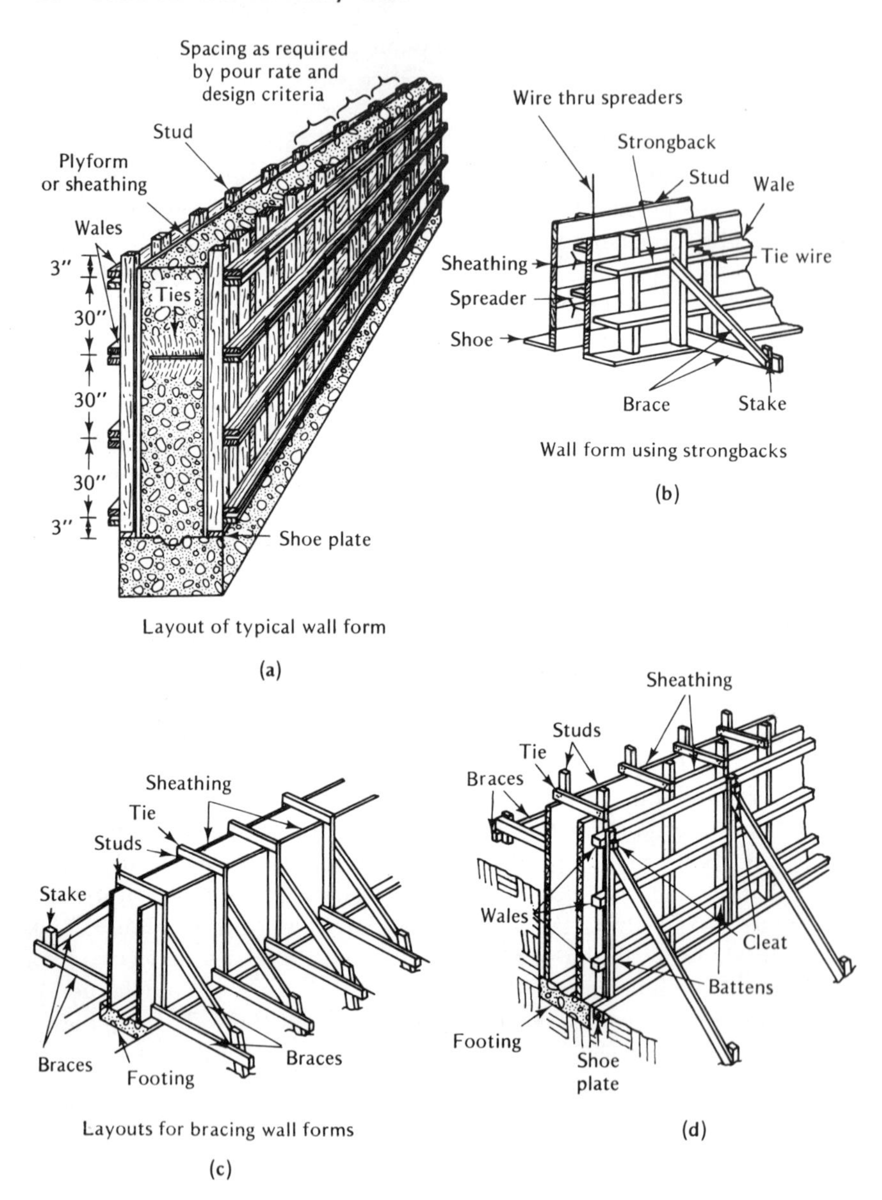

Figure 5–5 Types of Wooden Wall Forms

In Figure 5–5d battens are fastened to the wales of one member. The sole plate holds the battens in alignment, and cleats are used to aid in bracing. Notice that studs extend above the wall sheathing and ties are nailed across each.

TABLE 5–1 CONCRETE PRESSURES FOR COLUMN AND WALL FORMS (COURTESY OF AMERICAN PLYWOOD ASSOCIATION)

Pour Rate (ft./hr.)	Pressures of Vibrated Concrete (psf) (a), (b), (c)			
	50° F		70° F	
	Columns	Walls	Columns	Walls
1	330	330	280	280
2	510	510	410	410
3	690	690	540	540
4	870	870	660	660
5	1050	1050	790	790
6	1230	1230	920	920
7	1410	1410	1050	1050
8	1590	1470	1180	1090
9	1770	1520	1310	1130
10	1950	1580	1440	1170

Notes: (a) Maximum pressure need not exceed 150h, where h is maximum height of pour.
(b) For non-vibrated concrete, pressures may be reduced 10%.
(c) Based on concrete with density of 150 pcf and 4 in. slump.

TABLE 5–2 ALLOWABLE PRESSURES ON PLYFORM WITH FACE GRAIN ACROSS SUPPORTS (PSF) (COURTESY OF AMERICAN PLYWOOD ASSOCIATION)

Support Spacing (inches)	PLYWOOD THICKNESS (inches)					
	½	⅝	¾	⅞	1	1⅛
4	3265	4095	5005	5225	5650	6290
8	970	1300	1650	2005	2175	2420
12	410	575	735	890	1190	1370
16	175	270	370	475	645	750
20	100	160	225	295	410	490
24			120	160	230	280
32					105	130
36						115

Plywood continuous across two or more spans.

TABLE 5–3a: MAXIMUM SPANS FOR JOINTS OR STUDS MADE OF DOUGLAS FIR-LARCH NO. 1 OR SOUTHERN PINE NO. 1, IN INCHES (WALES SEPARATION)

Equivalent uniform load (lb/ft)	Continuous over 2 or 3 supports (1 or 2 spans) Nominal size								Continuous over 4 or more supports (3 or more spans) Nominal Size							
	2x4	2x6	2x8	2x10	2x12	4x4	4x6	4x8	2x4	2x6	2x8	2x10	2x12	4x4	4x6	4x8
400	40	50	77	98	117	52	82	104	43	65	86	110	133	65	99	122
600	30	47	62	80	97	46	72	94	31	49	64	82	100	56	81	107
800	24	38	50	64	78	42	63	83	25	39	52	66	81	48	70	93
1000	21	33	43	55	67	39	56	74	21	34	44	57	69	41	63	83
1200	19	29	38	49	60	34	51	68	19	30	39	50	61	35	55	73
1400	17	27	35	45	54	30	47	62	17	27	36	46	56	31	49	64
1600	16	25	32	41	50	27	43	56	16	25	33	42	52	28	44	58
1800	15	23	30	39	47	25	39	52	15	24	31	40	48	26	40	53
2000	14	22	29	37	45	23	36	48	14	22	29	38	46	24	37	49
2200	13	21	28	35	43	22	34	45	14	21	28	36	44	22	35	46
2400	13	20	26	34	41	20	32	42	13	20	27	34	42	21	33	44
2600	12	19	26	33	40	19	31	40	13	20	26	33	40	20	31	41
2800	12	19	25	32	38	19	29	38	12	19	25	32	39	19	30	39
3000	12	18	24	31	37	18	28	37	12	19	24	31	38	18	29	38
3200	11	18	23	30	36	17	27	35	12	18	23	30	37	18	28	36
3400	11	17	23	29	36	17	26	34	11	18	23	30	36	17	27	35
3600	11	17	22	29	35	16	25	33	11	17	23	29	35	16	26	34
3800	11	17	22	28	34	16	24	32	11	17	22	29	34	16	25	33
4000	10	16	22	28	34	15	24	31	11	17	22	28	34	15	24	32
4500	10	16	21	27	32	14	22	29	10	16	21	27	33	14	23	30
5000	10	15	20	26	31	13	21	28	10	16	20	26	32	14	22	28

*Spans are based on PS-20 lumber sizes. Single member stresses were multiplied by a 1.25 duration-of-load factor for 7-day loads. Deflection limited to 1/360th of the span with $\frac{1}{4}''$ maximum. Spans are center-to-center of the supports.

TABLE 5–3b: MAXIMUM SPANS FOR DOUBLE WALES MADE OF GRADE NO. 1, IN INCHES (TIES SEPARATION)

Equivalent Uniform Load	Continuous over 2 or 3 supports (1 or 2 spans)									Continuous over 4 or more supports (3 or more spans)								
	Nominal size									Nominal size								
(lb/ft)	2 x 4	2 x 6	2 x 8	3 x 4	3 x 6	3 x 8	4 x 4	4 x 6	4 x 8	2 x 4	2 x 6	2 x 8	3 x 4	3 x 6	3 x 8	4 x 4	4 x 6	4 x 8
1000	35	52	69	44	67	89	49	76	98	36	56	74	52	75	99	60	89	115
1200	30	47	62	41	61	81	46	72	94	31	49	64	47	69	91	56	81	107
1400	27	42	56	39	57	75	44	67	89	28	43	57	41	64	84	52	75	99
1600	24	38	50	36	53	70	42	63	83	25	39	52	37	58	77	48	70	93
1800	22	35	46	33	50	66	40	59	78	23	36	48	34	53	70	44	66	88
2000	21	33	43	30	47	62	39	56	74	21	34	44	31	49	64	41	63	83
2200	20	31	41	28	44	58	36	54	71	20	32	42	29	45	60	38	59	78
2400	19	29	38	26	41	54	34	51	68	19	30	39	27	43	56	35	55	73
2600	18	28	37	25	39	51	32	49	65	18	28	38	26	40	53	33	52	68
2800	17	27	35	23	37	49	30	47	62	17	27	36	24	38	50	31	49	64
3000	16	26	34	22	35	46	29	45	59	17	26	34	23	36	48	29	46	61
3200	16	25	32	21	34	44	27	43	56	16	25	33	22	35	46	28	44	58
3400	15	24	31	21	32	43	26	41	54	15	24	32	21	33	44	27	42	56
3600	15	23	30	20	31	41	25	39	52	15	24	31	20	32	42	26	40	53
3800	14	22	30	19	30	40	24	38	50	15	23	30	20	31	41	25	39	51
4000	14	22	29	19	29	38	23	36	48	14	22	29	19	30	39	24	37	49
4200	14	21	28	18	28	37	22	35	46	14	22	29	18	29	38	23	36	48
4400	13	21	28	17	27	36	22	34	45	14	21	28	18	28	37	22	35	46
4600	13	20	27	17	27	35	21	33	44	13	21	28	17	27	36	22	34	45
4800	13	20	26	17	26	34	20	32	42	13	20	27	17	27	35	21	33	44
5000	13	20	26	16	26	34	20	31	41	13	20	26	17	26	34	20	32	42

*Spans are based on Ps-20 lumber sizes. Single member stresses were multiplied by a 1.25 duration-of-load factor for 7-day loads. Deflection limited to 1/360th of the span with $\frac{1}{4}''$ maximum. Spans are center-to-center of the supports.

The location of the various members, studs, wales, ties, and braces are based on several complex formulas. These have been reduced to tabular form, where the pour rate at different concrete temperatures is provided (Table 5–1). With these data plywood thickness and stud separation can be selected (Table 5–2), and the wale and tie separation schedule can be obtained from tables similar to Table 5–3.

An example of the methodology follows:

1. Select a pour rate: 3 ft/hr.
2. Read Table 5–1: the pressure is 540 psf.
3. Read Table 5–2: pressure of 575 approximates the expected actual pressure of 540. (*Note:* Use ⅝-in. plywood with a stud spacing of 12 in. o.c.)
4. Interpolate Table 5–3a data for 2 × 4 continuous over four or more supports (studs) as 32 in. o.c. wale separation.
5. Interpolate tie separation from Table 5–3b as: 26 in. o.c. (e.g., 540 psf × 32/12 ft = 1440 lb/ft). From Table 5–3: 26 in. o.c. (interpolated).
6. Load on ties: load on wales × tie spacing, in feet: 1440 lb/ft. × 26/12 ft = 3120 lb.

To grasp the importance of pouring the form, look at Table 5–4. If the form was designed for a 3-ft/hr pour rate at 70°F, an attempt to increase the pour rate to 5 ft/hr results in a lateral pressure increase of over 900 lb/ft^2. This is a 56 percent gain in pressure to all elements in the form. Had 3500-lb ties been installed, the new pressure would be over 4500 lb, exceeding the tie strength by 1000 lb. Other parts of the form, such as wale separation and stud separation would also be inadequate and structural collapse would occur.

3. *Location and size of bracing:* The primary purpose of bracing is to establish and maintain form accuracy. Therefore, bracing is installed at acute angles of less than 45 degrees as Figure 5–5b, c and d shows. This places the strongest portions of the bracing along the vertical plane and distributes the forces back and down toward the stake. As long as the angle is as shown, form lifting does not occur, wind forces at the face of the forms do not affect form position, and plastic concrete's weight and lateral force are contained within the necessary confines.

Normally, nominal-2-in. stock is used for bracing. These should be 2 × 4s or 2 × 6s; where false work or a runway is needed, use 2 × 8s. The strong back shown in Figure 5–5b is usually made from

TABLE 5–4 LATERAL CONCRETE PRESSURES FOR VARIOUS TEMPERATURES (COURTESY OF AMERICAN PLYWOOD ASSOCIATION)

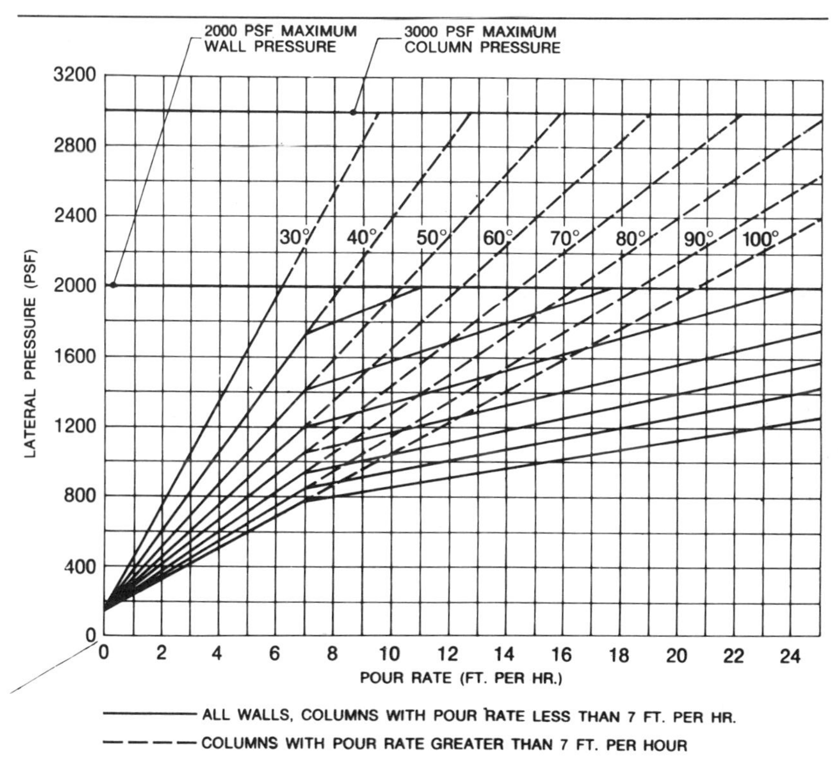

4 × 4s or 4 × 6s, and angular braces for some narrow-depth walls may be made from nominal-1-in. stock.

Of particular interest is the way bracing is used at joints as well as at inside corners, since these three areas could produce unwanted errors. If the joints are constructed properly, the sheathing is backed by 2-in. stock or studs. If two segments are butted to form a joint, some locking device, nailing, or clamping, must be used. The surface against the concrete must have perfect edge alignment for all architectural work and should be minimized on other types of walls.

4. *Tolerances*[2]: Industry standards indicate several values of tolerances.

[2] *Recommended Practices for Concrete Formwork,* American Concrete Institute (ACI 347–68).

a. ¼ in. out of plumb up to 10 ft high.
b. ¼ in. out of level per 10 ft of run.
c. ½ in. variation of linear building line from established position in plan.
d. ¼ in. variation in wall openings.
e. —¼ to +½ in. variation in cross-sectional dimensions (thickness) of wall.

5. *Rate of pour:* One of the basic requirements for construction, the rate of pour, is used as a basis to determine size of members and position of members within the form. The rate of pour, together with the temperature, directly affects the total pounds per square foot of lateral pressure which plastic concrete exerts upon the form.

6. *Consistency of concrete:* The cement/water ratio and the established slump play important roles in form pressure. Both should be maintained to develop the prescribed cured strength.

Also of special concern is the need to select the proper size of coarse aggregate. Recall that its size must not exceed one third of the width of the form. But, in wall construction the aggregate must pass easily between reinforcement steel bars or wire and around planned orifices. Therefore, a smaller-diameter (e.g. ¾-in.) coarse aggregate may provide easier workability.

7. *Control of deflection:* Adequate bracing and proper separation of studs, wales, and ties control deflection for the most part. Caution must be exercised when casting the concrete into the form. Concentrations of the mixture may cause deflections of sheathing, stud, or bracing, resulting in bulges. If this happens, tolerances will be violated. But when concrete is properly placed, none of these problems exist.

8. *Architectural requirements:* Designs that are to be made a part of the finished wall are fastened to the form sheathing prior to assembly. Their position and character must be proper and they must be adequately secured. If designs require a through section, those forms should be made for easy removal, and their surfaces that come in contact with the concrete must be treated with oils or lubricants like other parts of the sheathing.

9. *Needs of related trades:* Offsets, throughways, anchor blocks, anchor rods, and the like are all part of the general requirements of form work. Each must be understood and, where needed, installed perfectly.

Many other, equally important requirements are needed for proper forms. They differ from the more general types in that they are used more selectively. Some are jointing two or more pours, making provisions for use of mechanical vibration tools, oiling some types of form sheathing, applying plastic sheathing over the form surface, and making walkways for men and dollies used for transporting the concrete.

General Method of Form Assembly

The first actual step in form assembly is to cut studs and sheathing panels (plywood or plyform). Subassemblies are made where studs are nailed to one sheet according to on-center stud requirements and holes are drilled for ties.

The shoe plate is either fastened to the footing or to the bottom of several panels and studs. In the former, the panel studs are toe-nailed to the shoe plate; in the latter the entire assembly is positioned on the footing and nailed in place.

Wales are installed, lower to upper, with ties and clamps. Additional panels are erected and more ties are used. Wales and panels continue to be installed until the desired length is reached or a corner is reached.

Temporary braces holding the form erect are now repositioned as the form is made plumb and in line with the foundation line.

The form's inner surface is prepared for concrete by oiling or covering as required, and opening forms, pipes, and the like are installed. Architectural adornments are installed now also.

Following the reinforcement steel installation, the opposite-wall form panels are treated, then erected and secured with their wales, clamps, and braces.

Last, corners are completed and braced. Final adjustments to plumb and alignment are made and walkway, ramp, and other false-work is added where needed. The form is now ready for the concrete.

THEORY OF POURING THE CONCRETE

The previous section clearly defined the need for determining a particular pour rate. All parameters of the form's construction are based upon this value. Further, the temperature plays a significant role. Look at Table 5–4 again and observe the effect variations in temperature have on the lateral pressure against the form. Using the example of 3 ft/hr used earlier, confirm that lateral pressure can vary from 400 psf at 100°F, to 1050 psf at 30°F. It therefore follows that temperature control must be considered during all pouring operations.

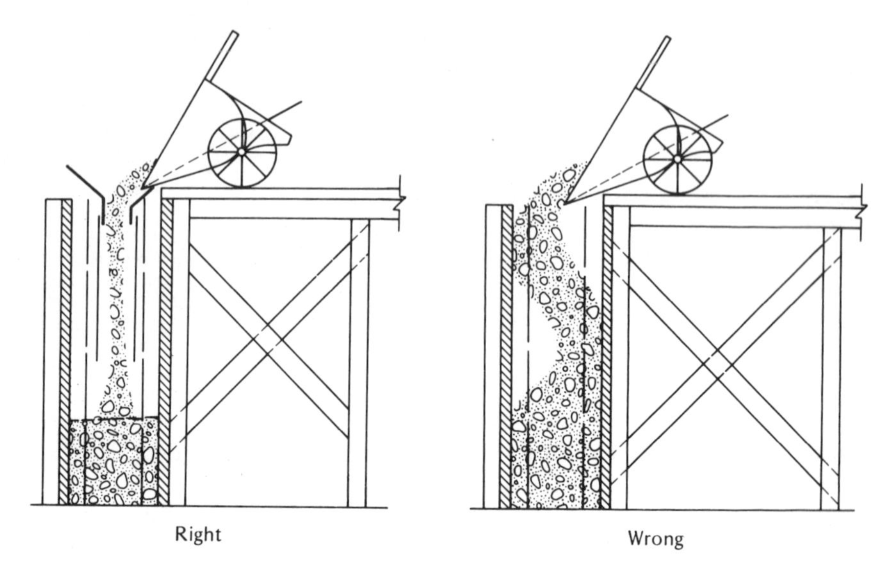

Placing concrete in top of form

Figure 5–6 Pouring Techniques

Pouring

Pouring the concrete into the wall forms should be done from *corner to center* and the casting should be done along the center of the wall's thickness not against either form surface. Bouncing the concrete from side to side within the form causes separation of large aggregate from the other components. Frequently, a down pipe is used to direct the mixture and avoid separation problems. Figure 5–6 shows the right and wrong ways to pour the mixture. Note the even distribution of large aggregate when done correctly.

Compacting

Compacting is essential so that the concrete mass encapsulates all steel, planned openings, architectural add-ons, and form-wall surfaces. This is accomplished by manual means (tamping) using nominal-1-in. or 2-in. stock or mechanically by vibrator-type tools.

The compacting causes the finer aggregate to fill rock holes and move toward the form walls, making for a finer finish of surfaces that are exposed when forms are removed. It is also needed to unite the separate layers of concrete that occur during the pouring operation.

The wall required in the examples used from Chapter 2 could effectively be compacted from above the form, since it is only 3½ ft high. For walls 7 or 8 ft high, for example, cutouts need to be made along the wall form to allow passage of the vibrator. These cutouts must be properly closed in once they have served their intended use.

Other Factors

Another factor to consider is the reinforcing metal's position while the form is being filled. It must remain as planned; therefore, care must be used that the concrete does not alter its position, and neither must the vibrator.

The concrete needs to be screeded flush with the form's top edge and may need to be floated. Also, anchor bolts must be installed while the mixture is in its plastic state.

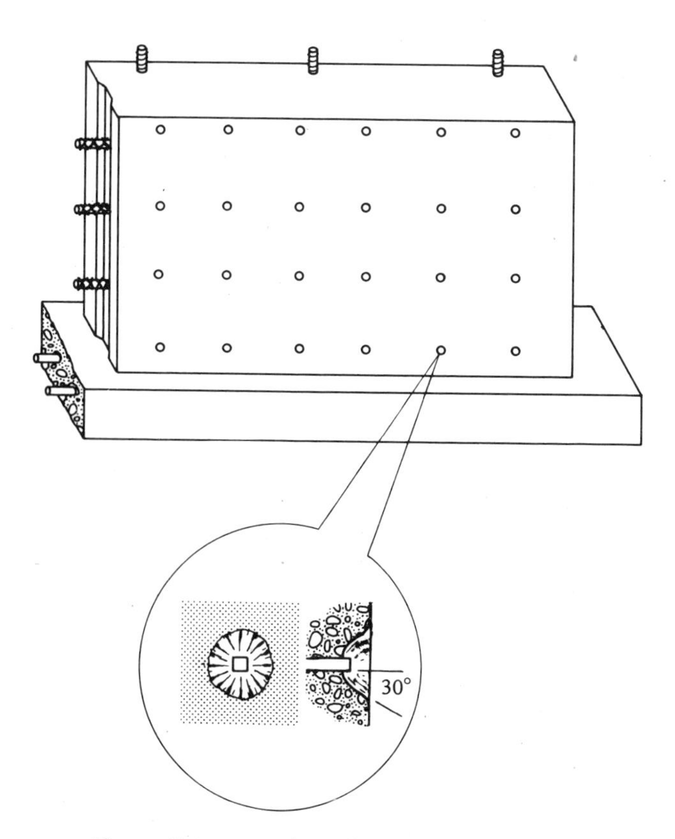

Figure 5–7 Poured Wall Ready for Finishing

Wall Finishing After Form Removal

Figure 5–7 shows a part of a poured wall with form removed and the ties snapped off below the surface. Grouting is used to fill these depressions as well as rock holes, should there be any. The wall surface to be in contact with backfilled earth must be waterproofed by using asphalt and in some cases, polyvinyl materials. Exposed surfaces are plastered and floated to the texture called for in the specifications.

INSPECTION OF THE CONCRETE WALL

Inspection	*Satisfactory/N.A.*	*Unsatisfactory*
Has proper alignment of the form been verified?	______	______
Has a final check been made of all clamps, braces, and ties?	______	______
Have provisions for pouring been completed?	______	______
Has a proper concrete mixture been ordered?	______	______
Have pouring operations been established?	______	______
Was pouring performed properly?	______	______
Was compacting performed?	______	______
Has the wall top surface been screeded or floated?	______	______
Have anchor bolts been installed?	______	______
Has allowable curing time been completed?	______	______
Have form ties been removed or snapped off below the wall surface?	______	______
Have grouting and plastering operations been completed?	______	______
Are all anchor bolts secure?	______	______

QUESTIONS

1. What is the importance of the following terms when building wall forms?
   a. Batten.
   b. Offsets.
   c. Plumb.
   d. Sheathing.
2. Name three types of ties.
3. State one reason why it is important for a mason to understand the principles of building forms.
4. What specifications does the carpenter extract from the elevation plan for use in building the form?
5. What are the nine general requirements for a wall form?
6. What does the selection of a pour rate have to do with placement of studs? Ties? Wales?
7. What is the maximum out-of-tolerance value (in inches) permitted in a form?
8. What would be some of the needs of related trades in forming a wall?
9. Generally outline the procedure used to construct a wall form.
10. What effect can atmospheric temperatures have on the lateral pressure on a form?
11. Why should bouncing the concrete mixture be avoided during the pouring phase?
12. What masonry tasks remain to be performed on a poured wall once the forms are removed?

Chapter 6

Concrete Slabs

Consolidation bringing together the concrete mixture from numerous placements so that no holes exist and the mass is unified.

Expansion joint a material placed within or a scoring of the concrete that allows it to expand without cracking.

Float a wooden tool used to finish a concrete surface.

Polyvinyl a plastic sheet, usually 6 mils thick, used under a slab as a moisture barrier; aids in insect control.

Screed a long, very straight board used for striking-off concrete.

Striking-off removing excess concrete to a level needed.

Tamp the compacting of concrete, using rakes or short lengths of lumber.

Trowel a steel tool with a flat surface which causes a concrete surface to become very smooth.

Wire mesh any of a variety of types of bonded wire forming a mat used to reinforce slabs of concrete.

OBJECTIVES—INTRODUCTION

Throughout the study of concrete's uses, the theme has been to determine what data are available from the blueprints and specifications that instruct in the construction of necessary forms and masonry work. Study of concrete slabs continues this theme, with the objectives: *to be able to translate a blueprint slab's detail into actuality,* and *to understand related trade requirements prior to pouring concrete.*

Three representative examples are used for the study: the patio floor, the basement floor, and the main dwelling floor. Each of the three requires different considerations and techniques, and yet there are similarities.

READING AND UNDERSTANDING THE SLAB BLUEPRINT AND SPECIFICATION DATA

Patio Floor

Figure 6-1 illustrates two partial-floor-plan drawings; each has a patio. From these two examples, individual and common data about location and dimensions can be obtained.

In Figure 6–1a, for example, the patio is inserted in an interior corner of the exterior wall. Notice that there are two sliding-door exits from the house to the patio (W1-D, W2-D). For the builder the overall dimensions are given as 11 ft (3.35 m) long by 6 ft 8 in. (2 m) wide.

In contrast, Figure 6–1b illustrates a patio that extends away from the house wall and either side of the patio door. Its dimensions are 13 ft (3.9 m) long by 5 ft 4 in. (1.6 m) wide.

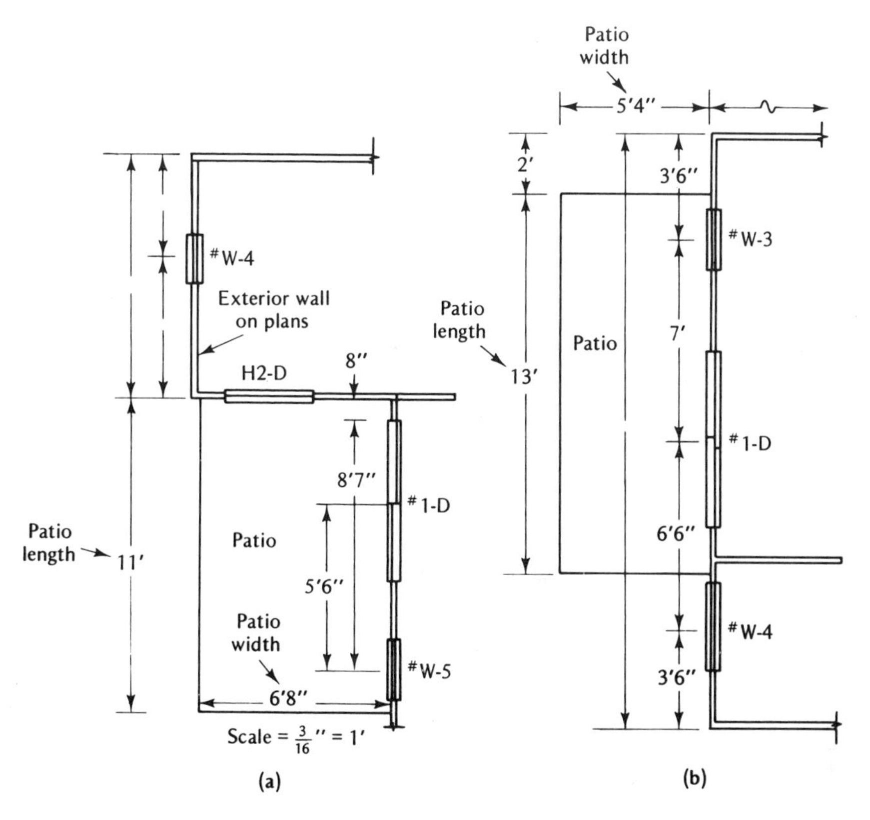

Figure 6–1 Patio Floor Plan

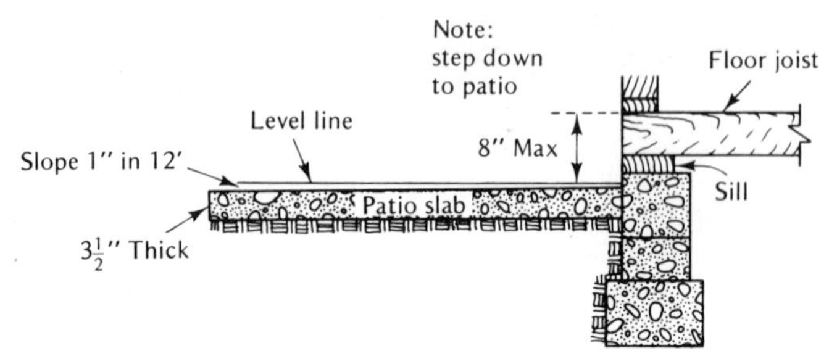

Figure 6–2 Patio—Elevation Details

Notice that the dimensions of the patios are shown close to the patio, within the patio (Figure 6–1a), or referenced from the exterior wall (Figure 6–1b).

From the elevation plan shown in Figure 6–2, more details about the patio and its slab are given. First, the amount of slope is given as *1 in. in 12 in.* This means that for each 12 ft of patio projecting from the house, the patio shall slope 1 in. This slope is needed to carry rainwater away from the house. Various patio widths result in different amounts of total slope. For example:

Patio width	*Total slope*
6 ft (1.8 m)	½ in. (1.25 cm)
8 ft (2.4 m)	11⁄16 in. (1.8 cm)
12 ft (3.6 m)	1 in. (2.5 cm)
16 ft (4.8 m)	1⅜ in. (3.4 cm)

If necessary, the slope can be increased to ¼ in./foot of run.

Second, the patio's beginning height is set at a *maximum* of 8 in. (20 cm) below the finished floor level of the house. It may be any dimension less than 8 in. If construction causes a height greater than 8 in., an intermediate step will be required. Finally, the slab thickness is given as 3½ in. (8.75 cm).

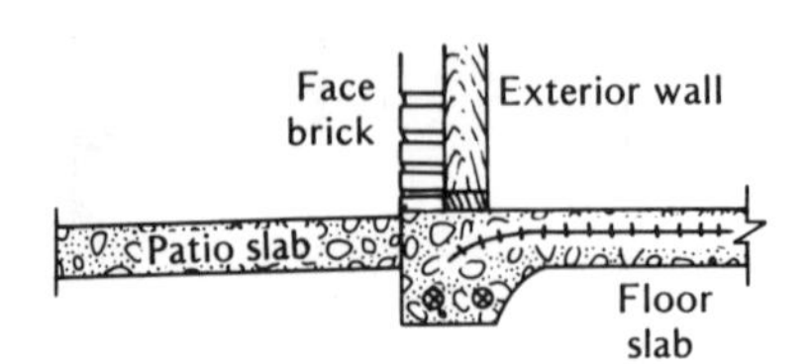

Figure 6–3 Patio Against a Slab Floor

Contrast Figure 6–3 with Figure 6–2 just examined and notice that where a slab concrete floor is used, the patio's floor height may be poured as close as 1 in. below the main floor level. This was not possible in the example because of the type of construction shown in Figure 6–2, where the siding on the wood frame extended below the sill.

Specification data would include the type of reinforcing wire mesh (e.g., "Patio floor to have 6 in. × 6 in. wire mesh."). Also listed may be the character of concrete (e.g., "Concrete for patio shall be cement/water ratio of 49–53, ¾-in. coarse aggregate, air-entrained agent, and 4 in. slump allowable with compressive strength equal to or greater than 2500 psf.").

Basement Floor

Rarely, if ever, is a floor plan given for a basement floor. However, the elevation plan always includes basement floor details. Figure 6–4 is a partial elevation drawing that shows foundation footing, wall, and basement floor. Observe that several details are given. First, the location of the floor on top of the footing. Next, the use of the expansion joint material between wall and floor, vinyl sheet material under slab and between footing wall and slab, and finally slab thickness as 4 in. (10 cm). Other details would be given in the specifications, which would be similar to those of the patio. Included would be type and quality of concrete and reinforcing wire.

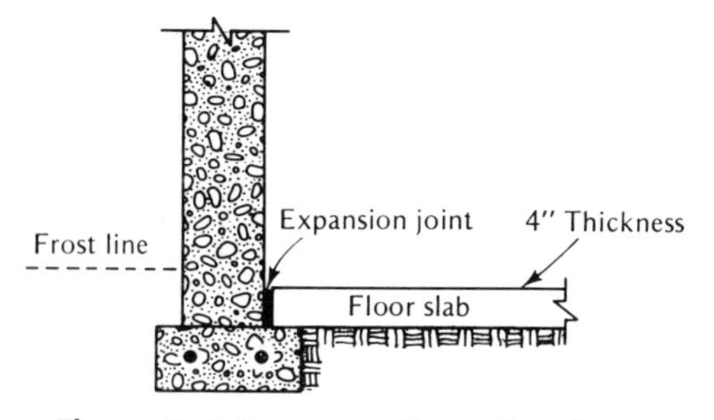

Figure 6–4 Basement Floor Detail

Main Dwelling Floor

An overview of a formed area such as the one shown in Figure 6–5 provides a total picture. It is the translation of the blueprints. The floor

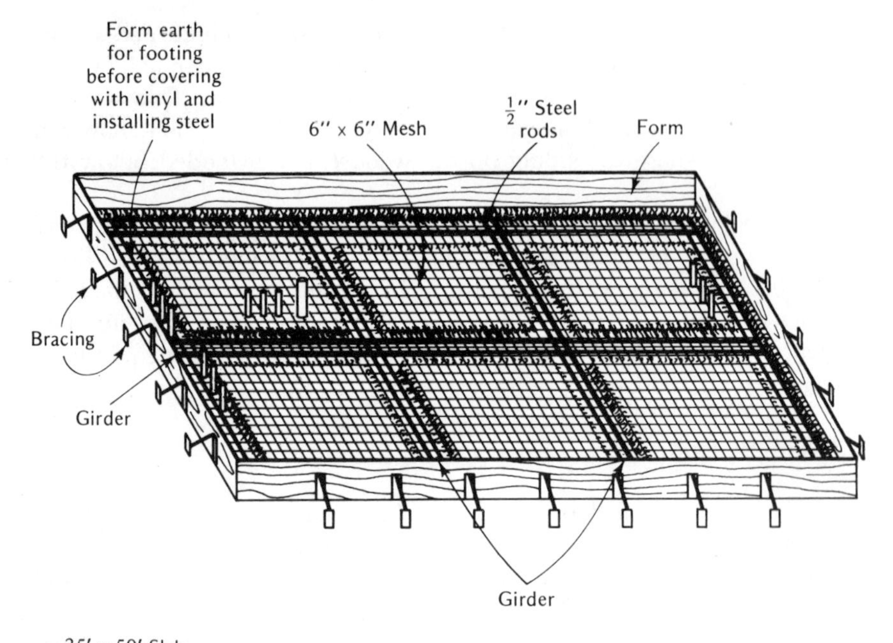

Figure 6–5 Overview of a Slab Formed

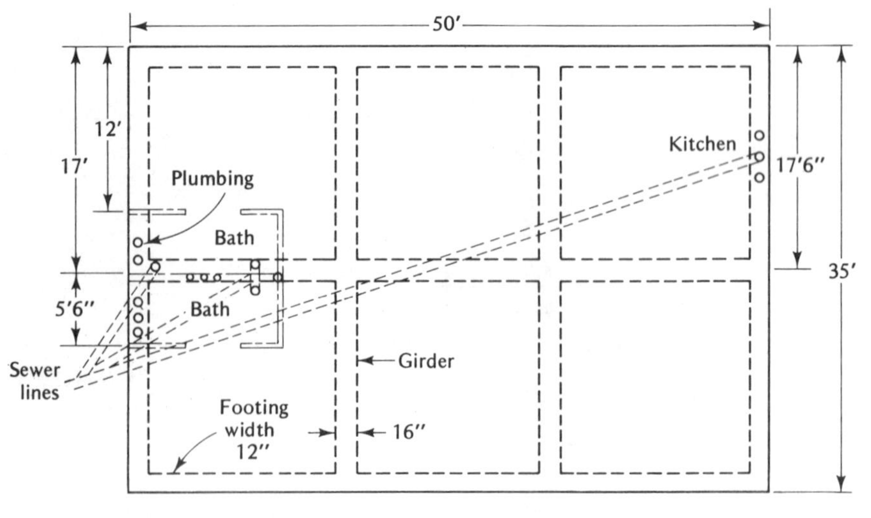

Figure 6–6 Floor Plan of a Slab Floor

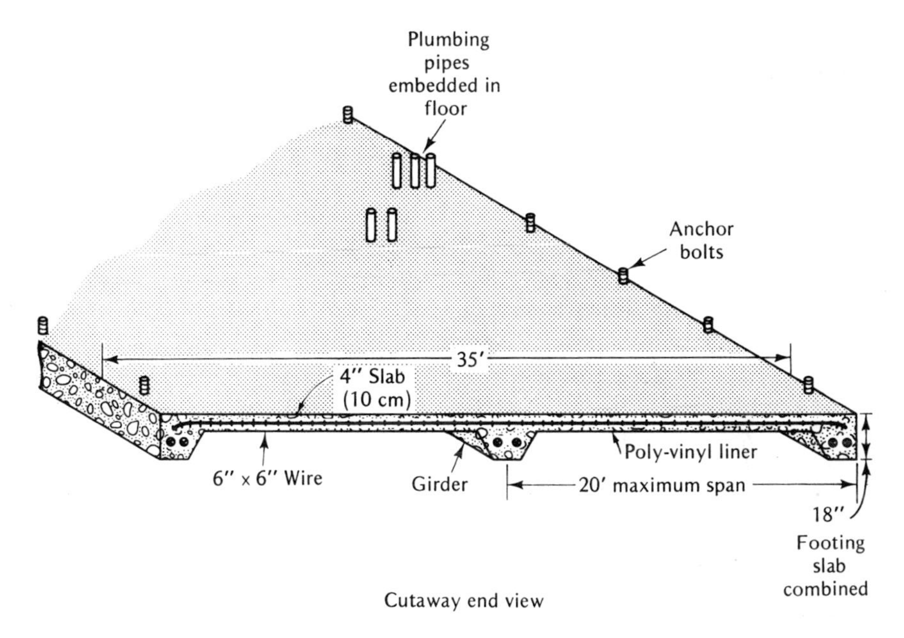

Figure 6–7 Elevation Data of Slab

plan provides dimensional data for the size of floor, and plumbing location data so that the plumber may install piping that needs to be located under the slab. Figure 6–6 is such a floor plan. Notice the correlation to the detail in Figure 6–5. Girder locations are provided as well as footing width. With the overlay of bathroom and kitchen plumbing data, the plumber locates the precise points where water and sewer pipes are to exit the concrete.

Elevation data may include a cutaway view of the slab, girder, and footing as Figure 6–7 shows, or it may be integrated into a full side or end elevation drawing of the building. In either case, closely observe the data provided. First, the footing and slab height are given as 18 in. combined. Second, steel rod and wire mesh data are shown. Third, girder placement is shown.

Normally the elevation details are restricted to the data listed in the previous paragraph and no projected view is given. Figure 6–7 has been modified from the standard elevation view so that two items can be shown. First, notice that the plumbing pipes extend above the poured concrete floor so that connections can be made later. Second, it shows the anchor bolts installed along the perimeter to provide sound anchoring for the exterior framed walls.

Not all slabs are poured at ground level. Where frost lines are several feet below ground level, a standard footing and foundation are required, as Figure 6–8 shows. After the foundation wall is raised to the desired height and has cured, earth is backfilled and tamped to the level needed. Plumbing is installed, if not done previously, and the concrete is poured.

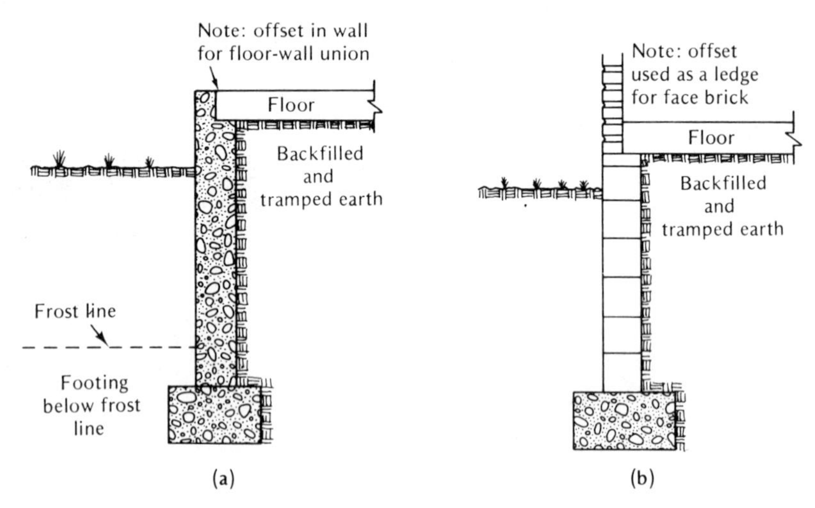

Figure 6–8 Elevations of Slabs above Ground Level

Notice that Figure 6–8a shows an offset on the inside of the foundation wall. This allows the floor, when poured, to rest on and join to the wall. Similarly, Figure 6–8b shows the floor overlapping part of the wall. But the foundation wall is fully below the slab floor. Implied here is that a form must be built to contain the plastic concrete.

THEORY OF FORMING FOR THE SLAB

Forming for poured slabs may or may not be required. For example, pouring the basement floor requires no forming; neither, for that matter, does the slab that is poured within the foundation.

However, there are several needs for forms. The patio slabs illustrated in Figures 6–1 and 6–2 need forms. These are made from 2 × 4 stock since the slab depths are only 3½ in. thick. These two examples show a clear need for exterior or perimeter forms.

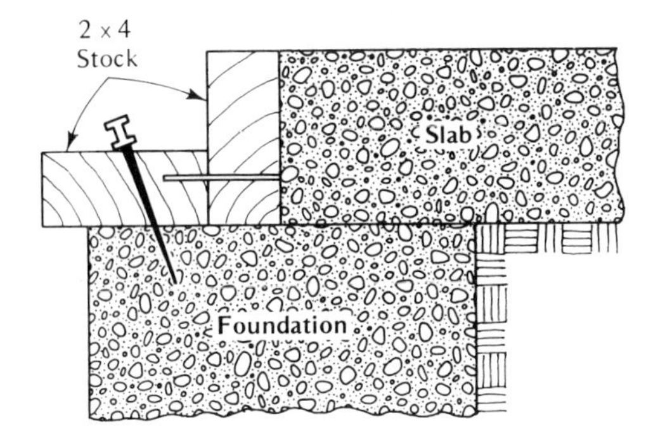

Figure 6–9 Slab Form for Recessed Slab Edge

Perimeter Forms

Refer again to Figure 6–5 (p. 84). It shows that an 18-in.-high form is needed to define the perimeter of this house. The 18 in. are needed for the 4-in. slab and 14-in. footing. Bracing consists of 1 × 3 or 2 × 2 pieces nailed to the form and stakes. Then backfill is placed against the form to keep it in place.

Where the floor is maintained back from the foundation edge as shown in Figure 6–8b (p. 86), a form is needed. Two 2 × 4s nailed together, then nailed to the top of the foundation wall as in Figure 6–9, make up the form.

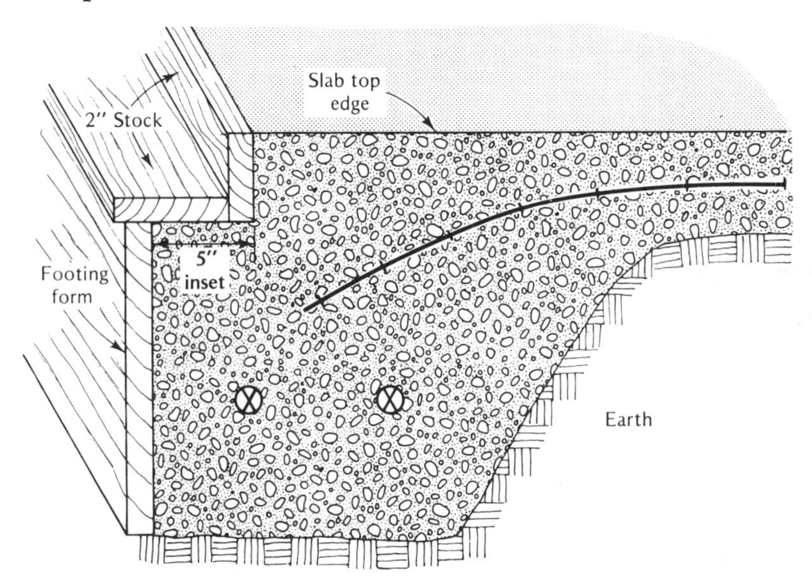

Figure 6–10 Offset in Pouring for Face Brick

Another example of a slab form (Figure 6–10) creates a depression where face bricks will later be laid. The footing form is installed in the usual manner. Then a 2-in. assembly is nailed on top of it so that the inward projection equals 5 in. The vertical member is as wide as the slab is thick.

Earth Forms

Also observe in Figure 6–10 that the inner wall of the footing is formed by the shape of the earth. This method should be used around the entire perimeter and to form the girders that bisect the area. The top of the earth should be measured and surfaced, 3½, 4, or more inches from the top of the form's edge to establish the thickness of slab.

THEORY OF POURING AND FINISHING CONCRETE

Pouring and finishing a concrete slab is a twofold effort. The first is the pouring, which includes preparation for pouring, and the second is the surface finishing, which includes applying curing techniques.

Preparation for Pouring

The preparation starts with laying in place of the polyvinyl sheeting that forms a moisture barrier. This should be done so that the various contours of girders and footing, if any, are clearly defined. Next the cutting, bending, and positioning of steel reinforcement rods is done. To assure that they will maintain their position, they may be tied with baling wire and supported with broken brick. Follow this activity with measuring, cutting, flattening, and positioning of the wire mesh.

Pouring the Concrete

Plastic concrete should be placed at the far end of the form, working back to the near side of the form. For the patio, for instance, start the placement near the house walls and work toward the outer form. Placing the concrete in this method assures even distribution of aggregate. Should other methods, such as *center dumping* or *near-to-far* placement, be used, large aggregate separates from the other components. The results would be uneven compressive strength, many rock pockets, and uneven distribution of air-entrained agents.

Movement and consolidation of the concrete is performed with shovel, garden rake, and hoe, so that all parts of the formed area are full. Although a vibrator can be used effectively to consolidate the concrete, it is not used on most small and residential concrete slabs. Instead, tamping with rakes and 2 × 4 stock is the standard consolidating method.

Screeding follows the placement phase and in this activity two men are needed, one for each end of the screed. A third man is desirable because he can pull excessive amounts of concrete or add needed amounts with a rake, hoe, or shovel.

The process shown in Figure 6–11 is as follows. The screed board, a long, very straight, nominal-2-in. member is placed across two parts of the form. Then a back-and-forth sawing motion is used by the men guiding the screed, together with a forward movement. The results are

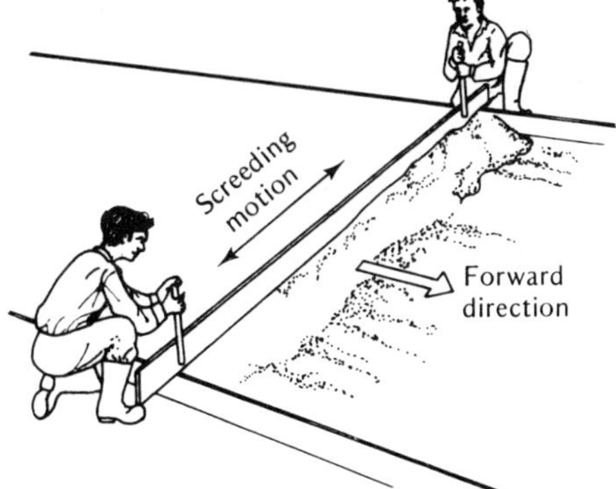

Figure 6–11 Screeding *(Courtesy of Portland Cement Association)*

a *striking-off* of the concrete's surface flush with the form height. It is desirable to screed the surface more than one time, and it should definitely be done just after the concrete has been poured.

Finishing Techniques

Floating, brooming, or troweling are methods used to finish the slab surface once screeding is completed. These activities are delayed, however, until the concrete has set to the point where it supports a person's weight with only slight indentation.

Floating requires the use of a wooden float, small as in Figure 6–12a, or larger as in Figure 6–12b. The hand float is used by moving it in sweeping arcs from a kneeling position. In contrast, the long-handled wood float allows floating to be done from outside the form.

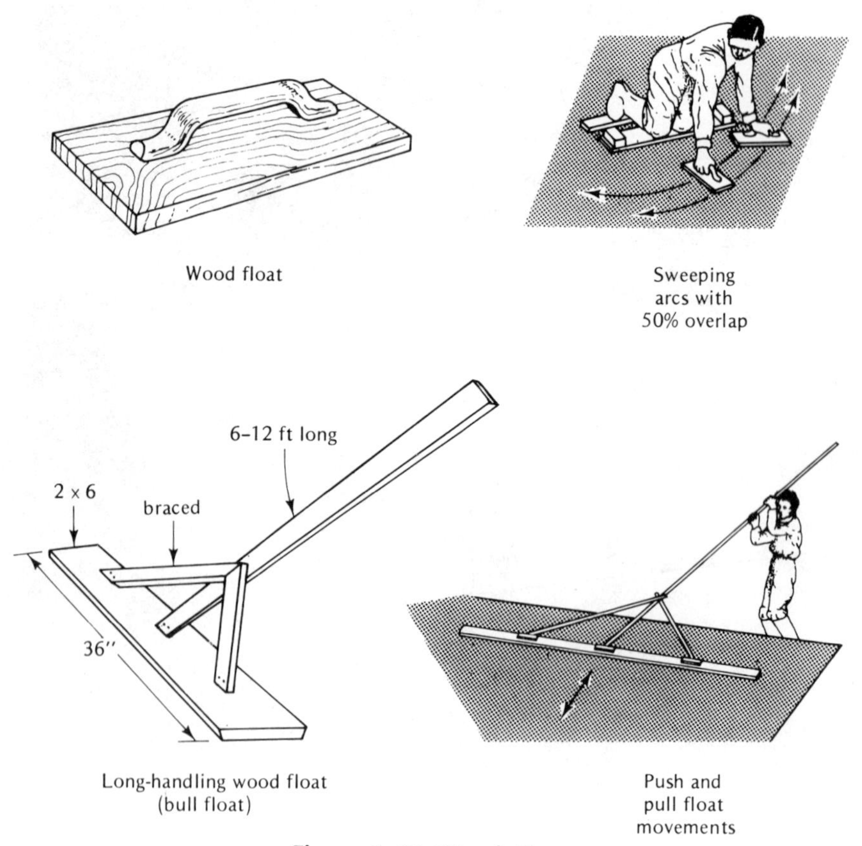

Figure 6–12 Wood Float

Several objectives and cautions must be kept in mind while floating. The purpose of floating is to embed aggregate particles just below the surface. It also provides an opportunity to remove slight imperfections, high or low spots, and to compact the concrete near the surface in preparation for other finishing operations.

Overworking the concrete with the float should be avoided, since this brings an excess of water and cement to the surface. These materials will form a thin weak layer that will wear away easily or flake off. And a final caution: at no time should cement be added to the surface; instead allow some time to elapse before continuing the floating.

Brooming may follow floating where the final texture of the slab is to show this finishing technique and produce a nonskid surface. Whereas floating is done in arcs, brooming is usually done in straight lines.

Various scoring depths can be achieved by selection of different brooms. A corn broom provides minimal scoring, usually 1/16 to 1/8 in. deep. The stiff bristle street broom may score up to 1/4 in., and steel bristle brooms may score even deeper. Or the surface may be troweled smooth with a steel trowel (Figure 6–13). The steel trowel produces a dense, smooth finish to the concrete surface. This technique

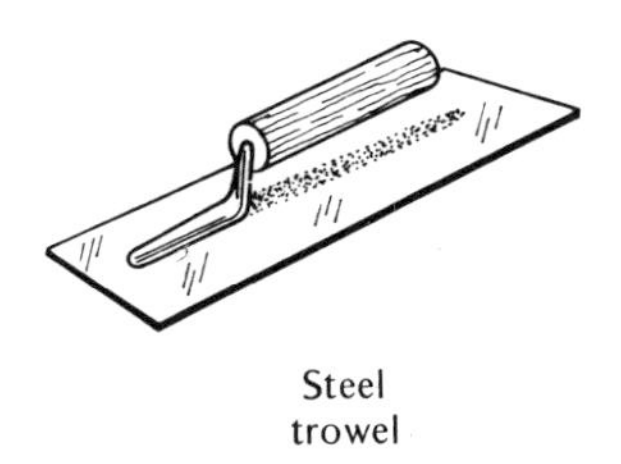

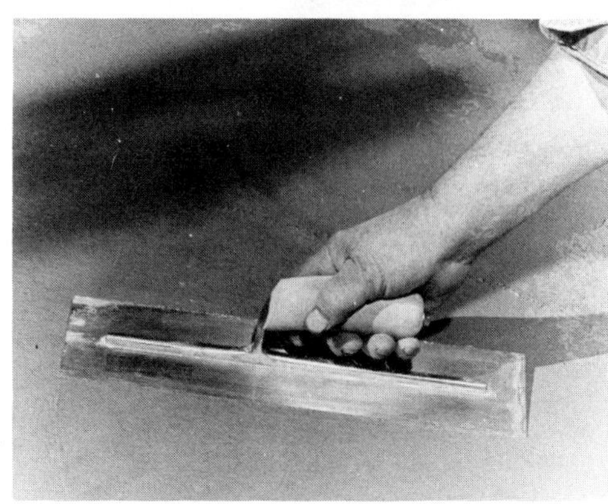

Figure 6–13 Trowel *(Courtesy of Marshalltown Tool Co.)*

begins after floating is completed and after the moisture film or sheen from floating has disappeared. As with hand floating, troweling is done with sweeping motions. With each pass the trowel's leading edge is raised slightly to smooth the surface and to prevent its digging into the surface.

Excessive troweling, like excessive floating, can create problems, such as hairline cracks, caused by a concentration of water and fine components. These cracks are further aggravated by too rapid drying or by cold temperatures. Fortunately, they can be eliminated by pounding the surface with the trowel, disturbing the set, and recombing the mixture.

Special Considerations

Slabs that form interior floor surfaces and others that will support columns or posts need *anchoring* devices installed prior to the setting of the concrete. These need to be carefully placed for on-center positioning as well as depth. Omitted or improperly spaced anchors cause a general reduction in unity between walls and floor.

Expansion joints may also be needed. If so, they may be cut into the surface during floating and troweling operations using tools as shown in Figure 6–14. The edger rounds the edges of the slab and the center-jointing tool is used to make expansion joints through the center of the slab.

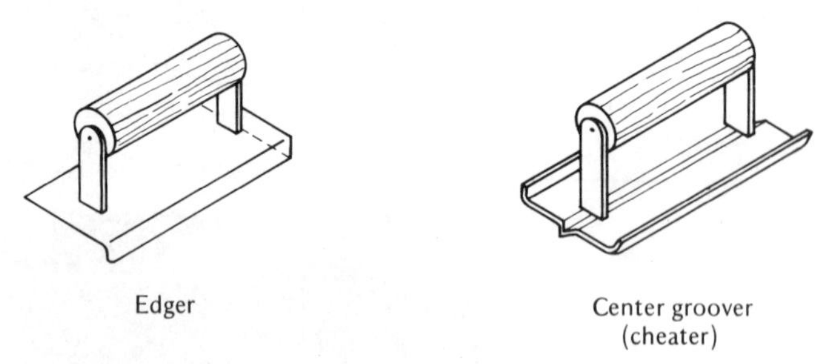

Figure 6–14 Joint-Making Tools *(Courtesy of Goldblatt Tool Co.)*

Curing Techniques

The concrete will also need to be cured. A variety of methods were detailed in Chapter 1. Best suited to patios and slabs in water sprinkling and covering with polyvinyl or waterproof paper. Where freezing weather is anticipated, use straw.

INSPECTION OF THE SLAB

Inspection	*Satisfactory/N.A.*	*Unsatisfactory*
Are forms adequate, soundly braced, properly aligned, and level or sloped as needed?	______	______
Has earth below the slab area been properly settled, contoured, and packed?	______	______
Has polyvinyl sheeting been installed over the earth?	______	______
Are all related trade requirements complete?	______	______
Have steel rods been installed?	______	______
Has wire mesh been cut and installed?	______	______
Was the concrete placed within the form correctly?	______	______
Have compacting and striking been completed?	______	______
Has floating been done properly?	______	______
Is the finished texture in accordance with plan requirements?	______	______
Have edges and center expansion joints been made properly?	______	______
Are anchor bolts in place?	______	______
Has curing been performed properly: For 7 days? For 28 days?	______	______
Have forms been removed?	______	______

QUESTIONS

1. Define the following terms.
   a. Consolidation.
   b. Expansion joint.
   c. Striking-off.
   d. Wire mesh.
2. Which plan provides the location data for slab or patio?
3. Which plan provides height, thickness, and sloping data about a slab?
4. Cite a complete specification for a slab made of concrete that meets industry requirements for quality and strength.
5. Should an expansion material be used between the basement wall and the floor?
6. Can a girder be cast when a floor is being poured?
7. Where would girders usually be included in a slab?
8. What are some requirements needed to prepare an area for a slab floor where the footing and part of the foundation are below ground?
9. Explain how a brick offset is formed with wood.
10. Name the two parts of pouring and finishing a slab.
11. Why should you fill a slab-formed area from the far to the near sides?
12. When should floating be started?
13. What effect does brooming create?
14. How can you avoid hairline cracks in the concrete's surface?
15. Name two methods that can be used to aid in curing concrete.

Chapter 7

Concrete Walks and Driveways

Backfilling the process of piling earth against the outer surface of a form.
Control joint a crease made in concrete that aids in expansion and minimizes cracking.
Edging the process of rounding the edge of freshly poured concrete; one of several finishing techniques.
Felt, 15-lb bituminous-impregnated paper used for many building purposes and for making strips used in imitation-flagstone construction.
Plot a segment of ground specifically located, having precise dimensions, and registered on official records.
Rock salt large crystals or fragments of common salt.
Screeding a process of striking-off concrete's surface with a screed tool or board.
Subgrade the grade below a slab of concrete, sidewalk, or driveway.
Utilitarian functional, useful, lacking beauty or esthetic appearance, as a sidewalk or driveway.

OBJECTIVES—INTRODUCTION

Concrete sidewalks and driveways play a significant role in landscaping the home's surroundings as well as provide sound, comfortable means for entering and leaving the home. They could be just a narrow strip of concrete laid for utilitarian purposes, but they also could be designed and integrated with shrubbery. In either case the plot plan includes their location. The elevation plan or a detail plan includes slopes, height positioning measurements, and other related data. The specifications include data on both sidewalks and driveways. This leads then to a single most important objective: *to translate into being a blueprint and specifications for a sidewalk or driveway.*

Included, of course, are the many tasks needed for building forms, grading, pouring, finishing, and curing the concrete. Forms are simple structures properly located and braced. Grading includes obtaining a desired depth from the top of the form. Pouring from ready-mix trucks is standard because many cubic yards or meters are usually needed. Finally, finishing and curing involves simple to extensive methods.

READING BLUEPRINTS AND SPECIFICATIONS TO LOCATE WALKS AND DRIVEWAYS

As a general rule, a set of plans is drawn to satisfy the requirements of the house. Only after a particular set of plans and a site are selected can a survey be made and a plot plan be drawn. The end result is to locate the house on the lot or plot (as designated in some locales).

So that the owner obtains an overview of the house location, driveway, sidewalks, trees, and shrubs, a rendition is drawn similar

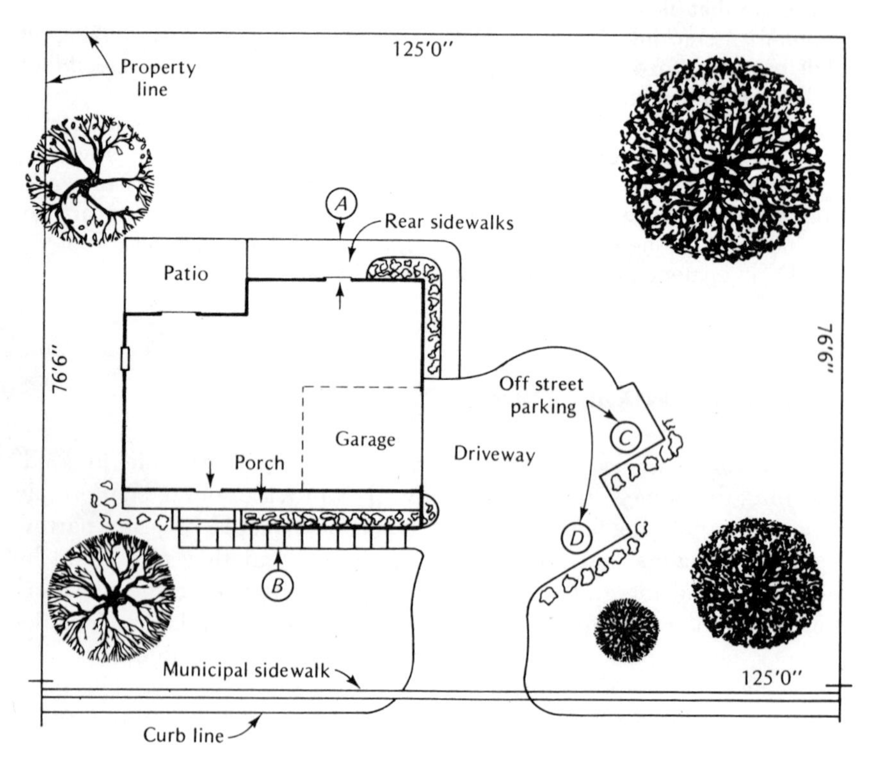

Figure 7–1 Overview of Plot Plan

to Figure 7–1. Even though this drawing lacks specificity for actual construction needs, it does provide valuable information for all interested parties.

Plot Plan with Landscape Data

With the house located properly, the owner and architect or builder design and agree on size, location, and design of sidewalks and driveways. If a strictly utilitarian approach is used, minimal design is involved. Walks are made as straight and as short as possible. Their finish will likely be broomed. Driveways are also simple straight lines and of minimal thickness. In contrast, where an integration of sidewalks and driveways with landscaping is planned, many ideas can be explored, and a final solution equitable to owner and builder can be arrived at. This planning and its execution may cost very little more than the utilitarian approach, so it is worthwhile exploring.

Before examining the plot plan, several items of general nature are identified. The average sidewalk is 3½ in. thick (a nominal 2 × 4), equating to approximately 10 cm. It may range from 30 in. to 6 ft in width, with 4 ft as the standard. A 4-ft-wide walk provides sufficient width for two adults to walk comfortably side by side.

The driveway usually is a nominal-6-in. thick, although well-reinforced 4-in.-thick driveways are adequate. It should be 3 ft wider than the vehicle to be parked on it, or roughly 10 ft minimum (3 m). Two-car driveways should be 20 ft wide, but frequently the width is reduced to 18 ft as a cost-saving measure. Where a single-width driveway leads to a double garage, flairing must be used near the garage. Also, an adequate distance from the garage to the edge of concrete must be determined to allow for turning the vehicle. The driveway should be built so that its connecting edge at the street is approximately 2 in. above the gutter, and it should be graded for proper slope away from the house to allow for proper drainage.

EXAMINE THE ARTISTIC RENDERING OF THE PLOT PLAN

The details in Figure 7–1 approximate:

1. Location of the house from boundary lines.
2. Location of sidewalks, municipal and private.
3. Location and design of driveways.
4. Location of trees and shrubs.

The sidewalk requirements include a walk at the rear of the house (see A) that adjoins the patio, connects to the rear door near the laundry (not shown), and circles around the end of the house to blend with the driveway. Another sidewalk is to be installed at the front of the house (see B) starting at the base of the front steps and running parallel to the house, finally connecting with the driveway. The sidewalk near the street is municipal property and is installed by contractors, usually at a cost to the developer or municipality.

The driveway is actually a composite of offstreet parking for two cars (see C and D) and a sweeping drive using curves to offset the rectangular proportions of the house. Sufficient area must be surfaced to allow cars backing out of the garage to back to their right, then enter the street driving forward. Parking slots C and D require a space 10 ft wide by 18 ft long.

From this drawing, which is the result of the owner's desires and needs and an architect's rendering or builder's sketch, parameters are developed and listed. These are or should be either included in the drawing or listed in the specifications.

Detailed Plot Plan

To illustrate each walk and the driveway, Figure 7–2, which is a modification of Figure 7–1, includes, in pen notation, parameters needed and agreed to. The measurements are as follows:

1. The front sidewalk is 4 ft wide × 38 ft long (see B).
2. The driveway is 17 ft wide from the street sidewalk to the edge of the garage.
3. The right-hand side of offstreet parking area D is 18 ft long and the end width is 10 ft.
4. The right-hand side of offstreet parking area C is 12 ft long and its end width is 10 ft.
5. The concrete area in front of the garage extends 18 in. beyond door.
6. The total distance from the garage to the opposite end of the driveway is 34 ft.
7. The radius of arc is 14 ft from the point shown.
8. The rear sidewalk is 6 ft from the rear wall extending to the patio (left) and 3 ft right of the rear door.
9. The balance of the sidewalk is 36 in. wide and is held 3 ft from the building.

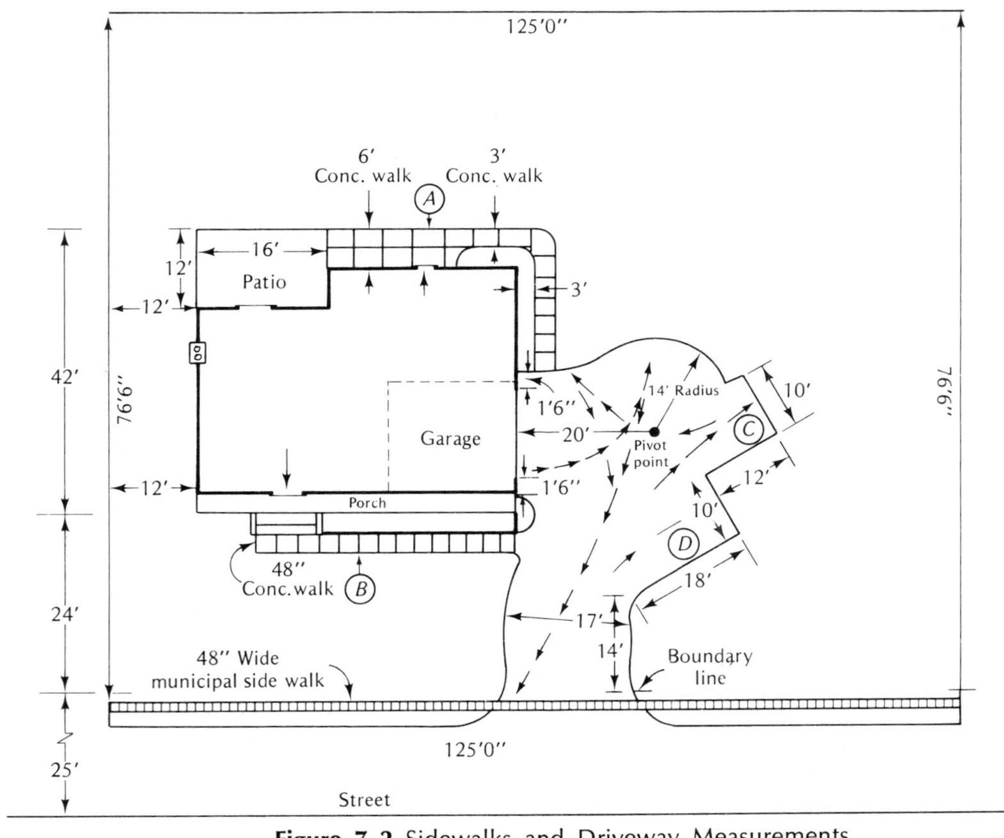

Figure 7–2 Sidewalks and Driveway Measurements

Considerable data are given that aid the builder; however, these are only dimensional and positional data. To be complete, elevation and specification details need to be provided.

Elevation Details

Figure 7–3 shows a typical elevation drawing that can be used by the builder/mason in locating the driveway where it attaches to the garage floor. It defines that the driveway is to attach 1½ in. below the garage floor level. The slope detail of ¼ in./ft is also given. The driveway is to be 6 in. thick and will be reinforced with 6 × 6 No. 10 wire mesh continuous throughout the driveway. The last detail provided is the requirement for compacted fill, graded. This means that earth moved from its original (virgin) state must be reestablished and firmed so that the concrete sets on a sound, solid base.

Sometimes it is difficult to compile a total picture of what is involved. Figure 7–4 may provide such a view as it shows the driveway from garage to street. The slope looks good, and notice that the driveway and sidewalk surfaces are flush. Note also that the driveway ends several inches above the gutter level.

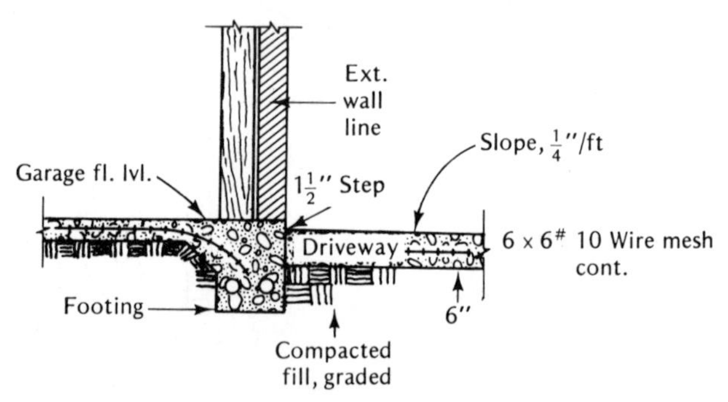

Figure 7–3 Elevation Garage—Drive Detail

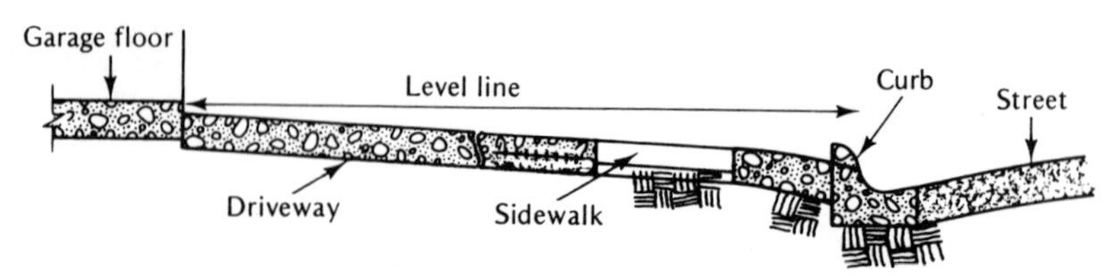

Figure 7–4 Cross-Sectional View of Driveway—Garage to Street

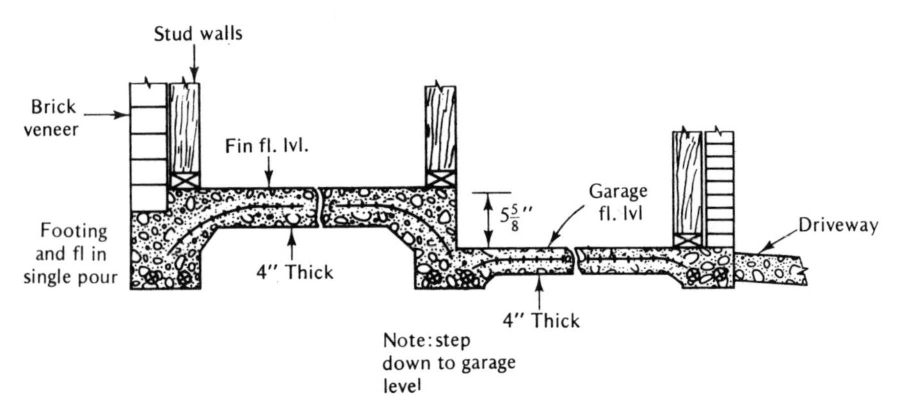

Figure 7–5 Levels of Surfaces Shown for Correlation

Next, the relationship of the driveway and sidewalk levels can be compared to the finished floor height by use of Figure 7–5. Starting at the left, the grade level is held at a minimum 8 in. below the finished floor level. The center of the drawing shows a 5⅝ in. (6.5 cm) step down to the garage floor level, and finally the drawing detail shown previously in Figure 7–3.

Specifications

Since so many details needed for the driveway are given in the plot plan and elevation details, few specifications are needed. However, the quality, type, and color of concrete to be used is probably listed. The type of finish also needs to be defined so that no misunderstanding develops.

Sidewalks are detailed for position and dimensions on the plot plan. Like the driveway, specification data are limited to grade, type, and color of concrete as well as type of finish. Sometimes sidewalk height needs to be listed but most often the walks are installed level with patios and driveways.

THEORY OF FORMING AND GROUND PREPARATION

The simplest of all forms built for concrete containment is the sidewalk and driveway. But before any forms can be installed, grading and form-site preparation must be completed. Then a uniform subgrade must be established after the forms are installed.

At the time that the sidewalks and driveway are to be installed, preliminary grading of the entire property makes installation easier. Contours of the grounds close to their final positions aid in establishing levels of walks and drive. The relative even surface also makes working more comfortable.

Since the soil has been disturbed, it does not have its original compactness, and therefore does not provide the solid, sound basis needed for concrete to lie upon. Subgrade soils within the form must therefore be prepared. Sand or a sand/pea gravel mixture is frequently used as fill and compacted with water spray and tamping.

Forming the Walk or Driveway

Forms for sidewalks are made from very straight 2 × 4s, metal forms, and plyform strips. The 2 × 4s and metal forms are used for straight runs and the plyform strips are used for curved segments. These form members are held in place with stakes and braces.

The process of installing a form follows a general pattern illustrated with the aid of Figure 7–6. First, the sidewalk must be located. These data are taken from the plot plan. Next, grade levels are established. These may be stakes driven at intervals of 8, 10, 15, or 20 ft and marked for finished concrete surface height. A transit tool is often used for this job. With the sides of the form located, a line (a mason's line) is strung from one point to another where one form will be installed. Then earth is removed below the line as necessary, to the depth of concrete to be poured.

At this time either of two methods may be used to install the form member. One, stakes separated at 4-ft intervals are driven into the ground along the location line. Then the form is positioned against the stake, set for height, and nailed. The second method requires positioning the form member for height, then nailing from stake to form.

Each end of the form member must be set for proper height. Then it must be verified throughout its length before being nailed. Succeeding form members are installed in like manner, starting with butting to prior member installed, establishing height of its opposite end, and verifying its height throughout its length.

The opposite form member used to develop the width of sidewalk or driveway is easier to install. First it is located by using a *finished-width guide* (Figure 7–6). This guide is made locally and makes work easier because it is used to establish form-member separation and is used as a level board. By placing it at intervals along the first installed form and at a right angle, the opposite form member's location can easily be determined. Earth is removed along this line to allow for form placement. Next the form is installed by placing the guide as shown in the figure, butting the form member to the guide, and staking behind it. This process is repeated several times throughout the length of the form member. Then a spirit level is used (a 24-in. one will do) on top of the guide to establish the height of the member being installed. The form is nailed to the stake when it is level.

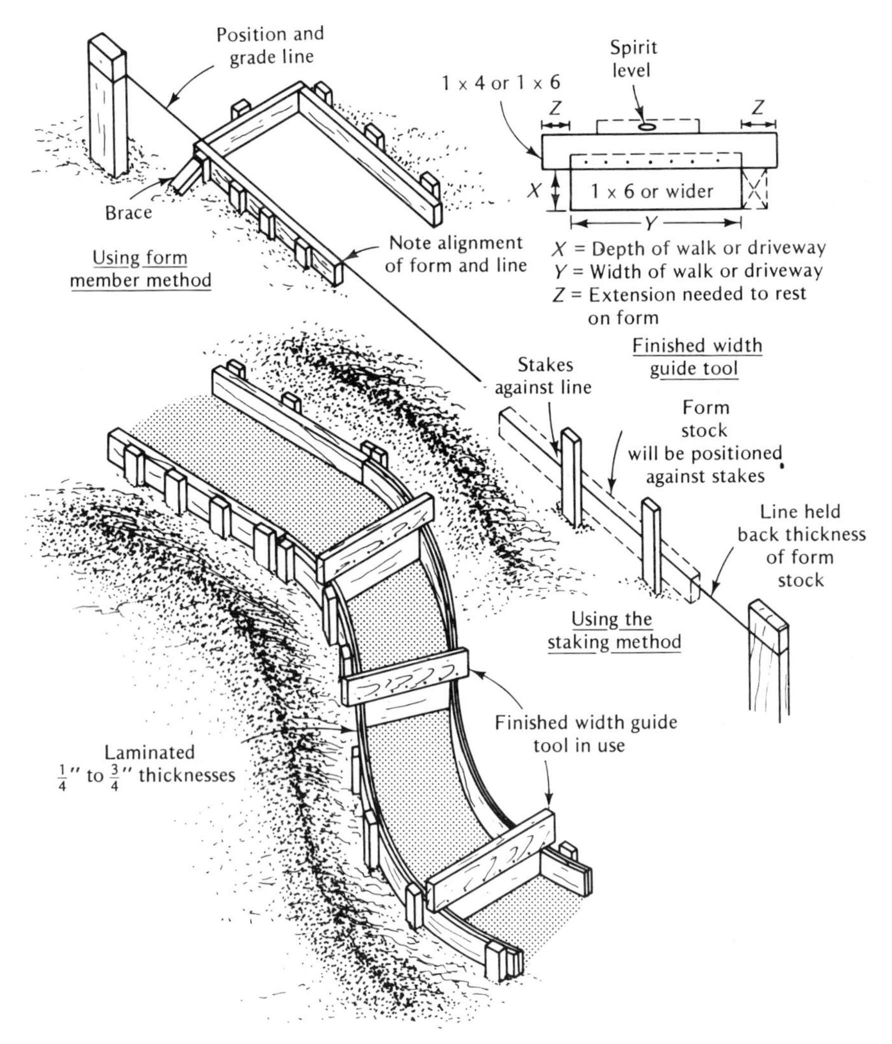

Figure 7–6 Forming Walks and Driveways

After all forms are in place braces are installed. There should be a brace each 6 to 8 ft or behind each stake. The criteria for staking follow several general rules: (1) where subgrade soils are loose and sandy, use braces at close intervals; (2) where soils are virgin and sound, braces may be spaced at greater intervals, since stakes are already soundly installed; and (3) where a concrete thickness of 5 in. or more is poured, use bracing to keep form walls perpendicular. After being driven in place, each brace is nailed to the stake it braces.

Return once again to the portion of Figure 7–6 that illustrates curved forms. Observe that the form materials are thin members. These may be made from ¼-in.-thick plyform, layered for strength, to ¾-in. stock that can be bent for gentle curves.

The procedures outlined for straight forms apply equally well to curved sections. Clearing soil below the form's position, followed by staking and nailing the form to its finished height, and bracing are all necessary steps. It is easy to keep the forms separated by use of the finished-width guide. This tool is usually used at the points where the greatest bends take place. Bracing is most important, because it not only holds the form perpendicular but reduces the forces caused by the bending of the form members and later the concrete's pressure.

Subgrading within the Form and Backfill

After the forms are installed, subgrading is performed. This consists of removing excess soil within the formed area, backfilling outside the form in areas that are lower than the form members, and filling in shallow areas within the form.

The *finished-width guide* is a real asset in grading. If its lower edge is built to the finished depth of the concrete, it can be used as a guide in grading. Throughout the length of the form, the tool can quickly and easily spot high and low areas. The high spots can be trimmed with a shovel and the low spots can be filled with soil, sand, or sand and gravel, and then tamped. Be sure, if installing granular fill, to wet it and tamp it so that it is soundly compacted. When all work has been done, preparation for pouring the concrete can begin.

THEORY OF POURING AND FINISHING CONCRETE

Pouring and finishing the sidewalk or driveway involves up to four separate tasks; installing reinforcing wire mesh, pouring the concrete, finishing the concrete surface with one of a variety of finishing techniques, and proper curing.

Installing Reinforcing Wire Mesh

Sidewalks may or may not need reinforcing. If the only traffic anticipated on the sidewalk is pedestrian or bicycle and subgrades are virgin or very well compacted, it may not be necessary. On the other hand, should cars or trucks periodically roll on its surface, reinforcing wire is needed.

The standard 6 × 6 No. 10 wire mesh is used most often for concrete sidewalk and driveway reinforcement. The wire needs to be unrolled and flattened prior to cutting it to fit. The cutting operation is easy if a pair of bolt cutters are used. Cut the wire several inches narrower than the width of the form. This prevents its being seen once the forms are removed. Finally, the cut wire is placed into the formed area, and the form is ready for pouring.

Pouring Concrete

Prior to placing the concrete, it is necessary to determine the quantity needed and the method of preparing it. For the quantity needed, use Table 7–1 as representative samples or compute the quantity needed as follows.

Step 1. Calculate the number of square feet within the formed area.

Step 2. Determine the concrete depth: 4, 5, or 6 in.
If 4 in. thick, allow *81 square feet* per yard of concrete.
If 5 in. thick, allow *64 square feet* per yard of concrete.
If 6 in. thick, allow *54 square feet* per yard of concrete.

Step 3. Divide total square feet by the factor indicated in step 2.

Example 1: sidewalk 4 ft × 40 ft × 4 in. thick
Step 1 4 ft × 40 ft = 160 ft^2.
Step 2 4-in.-thick slab; use factor 81.
Step 3 160 ft^2 ÷ 81 = 2 yards of concrete.

Example 2: driveway 18 ft wide × 26 ft long × 6 in. thick
Step 1 18 ft × 26 ft = 468 ft^2.
Step 2 6-in. thick; use factor 54.
Step 3 468 ft^2 ÷ 54 = 9 yards of concrete.

TABLE 7–1 CONCRETE NEEDS FOR SIDEWALKS AND DRIVEWAYS *

Thickness of Slab (in.)	*Number of Cubic Feet of Concrete for:* 10 ft^2	50 ft^2	100 ft^2	200 ft^2
4	0.12	0.62	1.23	2.47
5	0.15	0.77	1.54	3.09
6	0.19	0.93	1.85	3.70

* Based upon data from the Portland Cement Association, assuming a level grade and length × width. If spillage is anticipated, add 5–10 percent to the estimate.

It is possible to hand mix or rent a small mixer to mix the concrete for the sidewalk in Example 1. However, be aware that each cubic yard of concrete mixed means a movement (shoveling, hoeing, and the like) of 1 *ton* of materials, and for the sidewalk 2 tons of aggregate and cement will be handled by hand.

Generally, ready-mixed concrete brought by trucks is more practical. It is proportioned properly to meet compressive strength and water/cement ratios. Air-entrained agents are well mixed and the driver is available to assist (which they often do) not only in placing but in leveling (screeding).

The form is filled with the mixture, starting at the far end and working toward the near end. Recall that this method prevents separation of large aggregates from the mixture. As the concrete is being placed, do not allow it to pile up in the formed area; rather, move the truck's chute back and forth to distribute it evenly. The mass must be raked to all edges of the form and tamped thoroughly. Occasionally, the rake's teeth are used to lift the wire mesh into the concrete, but never to the surface.

After the placing is almost complete, the surface should be screeded twice. The excess should be moved to other parts of the form not presently filled. The filling, tamping, and screeding are continued until the entire form is full. Next, a bull float or darby is used to work fine aggregate and cement to the surface.

Finishing Concrete Surface

All finishing operations are begun after the water sheen is gone from the concrete's surface and the concrete can sustain foot pressure with

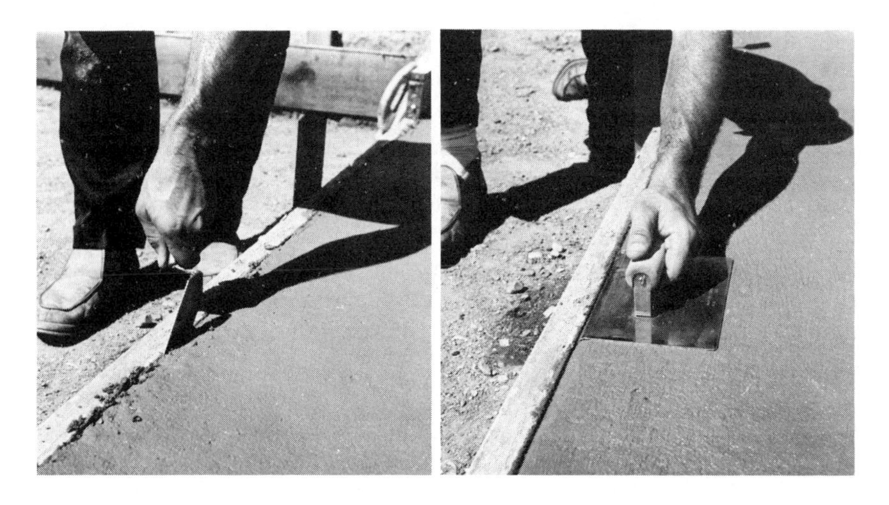

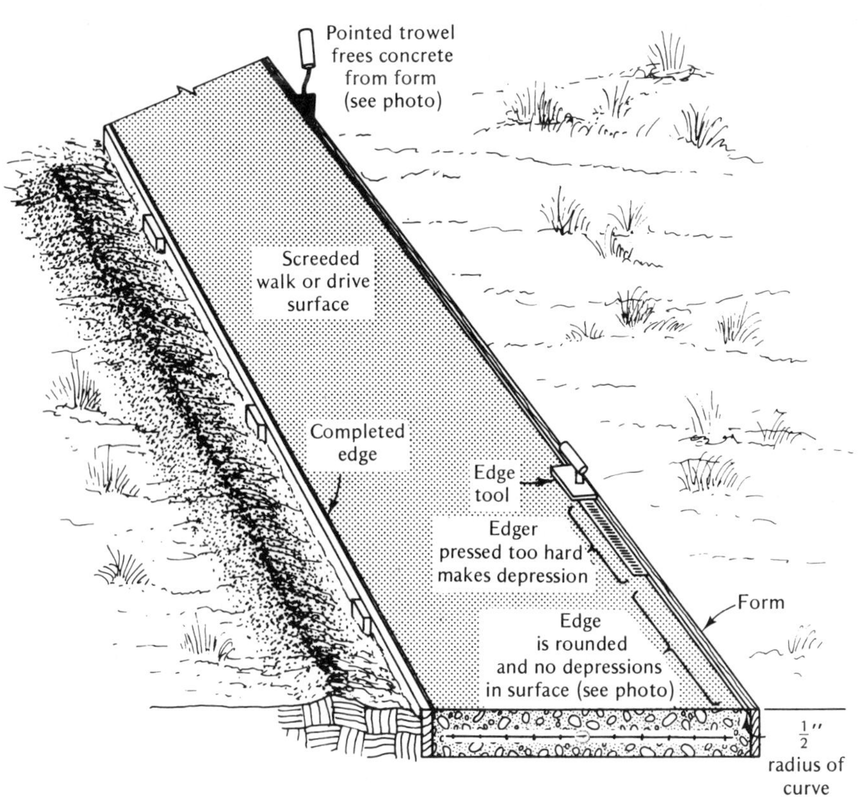

Figure 7–7 Edger Tool Operation *(Photo Courtesy of Portland Cement Association)*

only about ¼-in. indentation. The various finishing operations include edging, control jointing, floating, troweling, and brooming and other special effects.

Edging along the form is a simple operation. First, the pointed trowel is inserted between form and the concrete to cut the concrete away from the form. Next, the edge tool (Figure 7–7) is used to round the edge of the concrete. Edger tools with a ½-in.-radius rounded edge are recommended for walks and driveways. By using a forward and backward travel with the edger, the edge is rounded, and at the same time large aggregate is forced under the surface. This means that fine aggregate and cement are brought to the edge and surface.

Jointing operations using the control joint tool or groover follow edging. The groover (Figure 7–8) makes a depression in the concrete approximately one fifth of the thickness of the concrete. This depth is needed if the joint is to be effective, because control jointing can eliminate random cracks that could result from weathering. Recall from Chapter 1 that expansion and contraction of concrete occurs under varying conditions of moisture and temperature.

Control joints should be spaced at intervals equal to the width of the sidewalk or driveway and should, of course, run perpendicular to the edge of the concrete. As an example, make the control joints on 48-in. centers for a 48-in. sidewalk.

There are two easy methods for laying out the line of a control joint. One uses a chalk line, the other uses a straightedge. When using

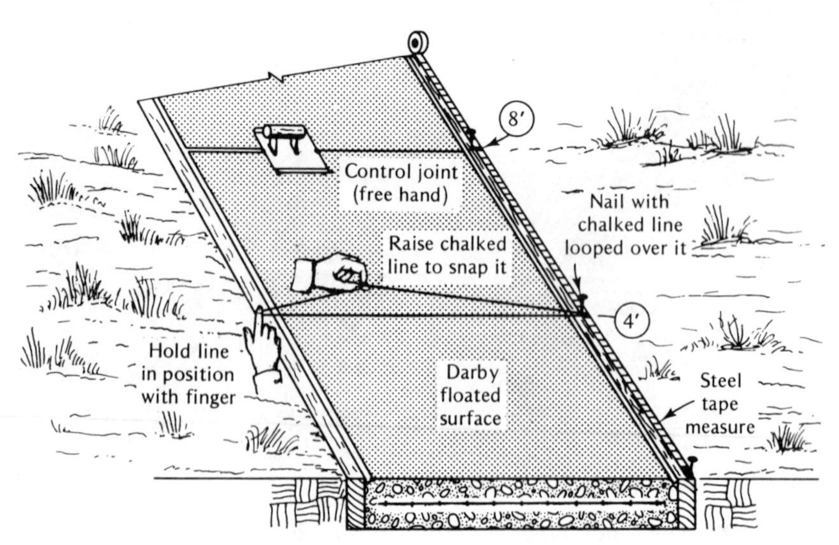

Figure 7–8 Control Joint Tool Operation

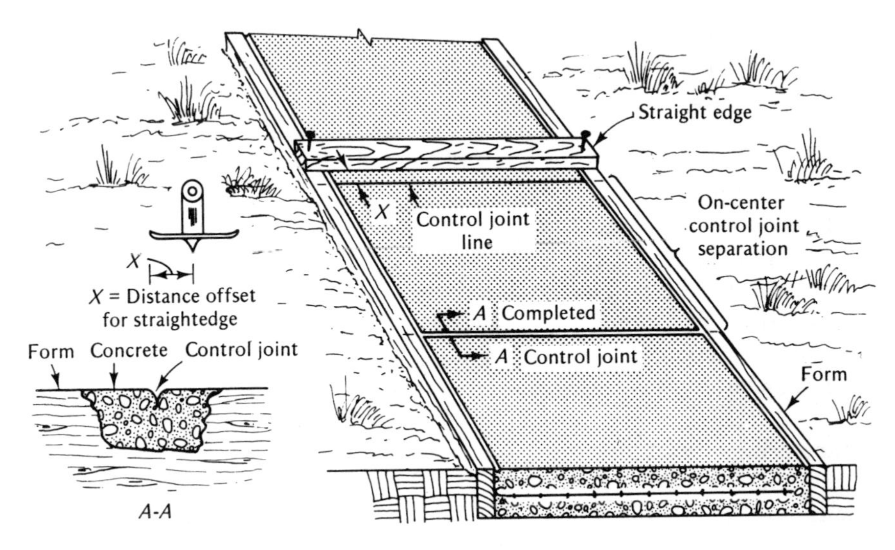

Figure 7–9 Using a Straightedge Guide

the chalk-line method, the position of each control joint is located by laying a steel tape measure along first one side of the form and then the other and marking off on-center spacings. Next a chalked line, nail and hammer, or a fellow worker are used to snap lines across the concrete. The jointing tool or a power saw [1] is used to cut the groove along the chalk line.

The second method requires the same process of locating joints but includes the use of a straightedge board. This board is set back from the joint equal in distance to the tool's width (Figure 7–9). As shown in the figure, the grooving tool is slid along the straightedge. If desired, a small nail can be used at both ends of the straightedge to secure its position. (*Note:* The straightedge is also effective when used with the power saw.)

The hand tool is started at the edge of the concrete by lifting the front edge, lowering the rear, forcing it down into the concrete, and pushing the tool away from the edge. This starting technique may be used several times until the tool reaches full depth. From there on, a forward and backward motion is used throughout the joint's length.

---

[1] Power saws may be used with abrasive or a diamond blade during the period 4 to 12 hr after pouring.

A final word needs to be added regarding edging and control joint operations. One must be especially careful *not* to cause depressions with the tools being used. Should their flat surfaces make depressions in the concrete surface, further finishing operations may be affected and the marks may be difficult to remove.

Floating operations follow jointing operations for three purposes: (1) they embed large aggregate, (2) they remove imperfections left from darbing and edging operations, and (3) they consolidate the mortar near the surface of the concrete. Floating was explained in Chapter 6. To summarize, floats, usually made from wood, are used with a sweeping movement, creating arcs as one swings an arm in circular fashion (see Figure 6–1).

A sidewalk can be considered complete after floating because the surface created is even but not smooth. Therefore, it gives good traction for walking and tires on cars, bikes, and the like. A well-floated surface does not have any distinguishing marks, but, if desired, it can be made to look like swirls or circles. The floating tool is used to create these appearances.

If a smooth surface texture is wanted, the *troweling* operation must follow the floating operation. A metal trowel is used in a manner similar to the float. Sweeping the tool in arcs with its leading edge slightly raised smoothes the surface to a polished look. This type of finish is dangerous for walks and driveways because it is very slippery when wet and can be the cause of serious accidents.

Special Finishing Techniques

Broom finish The simplest of all special finishing techniques used on sidewalks and driveways is the broom finish. Usually, a large, stiff bristle corn push broom is used. Its stiff bristles actually cut into the soft concrete surface, making depressions. The forward or pushing motion creates the greatest force and the deepest marking. The backward or pulling motion makes finer markings because the bristles glide along the surface of the concrete. If brooming is used as the final operation, troweling need not be done, although edging, control jointing, and floating must be done (Figure 7–10).

Exposed aggregate A very popular and attractive finish is one that exposes the large aggregate. This technique involves considerable knowledge of concrete handling. One procedure is as follows. Seeding with 1 to 1½ in. of aggregate over the entire surface is the first operation. Next, this aggregate is embedded into the soft surface of

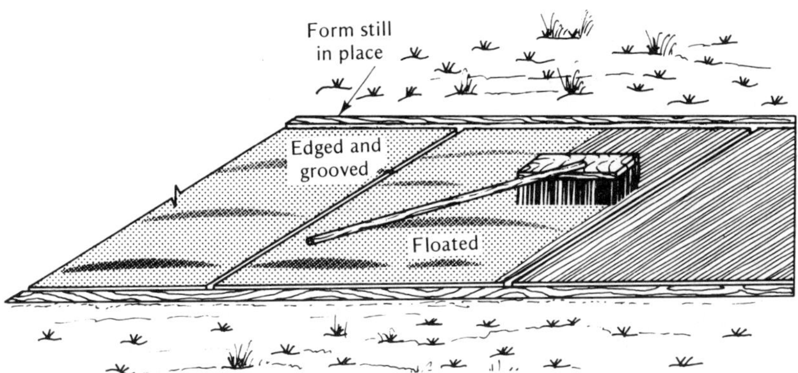

Figure 7–10 Broom Finish *(Photo Courtesy of Portland Cement Association)*

the concrete by tamping and bull floating. Excess mortar is removed from the surface and when the concrete can withstand a person's weight, a fine spray of water and a stiff, nylon-bristle push broom are used to expose the aggregate.

The same effect can be achieved without seeding by exposing the large aggregate contained within the concrete (Figure 7–11).

Rock-salt texture A unique texture can be obtained if after floating is finished, rock salt is sprinkled on and rolled into the surface. After the concrete dries, the salt is washed away using brooms and water, leaving pits and holes (Figure 7–12).

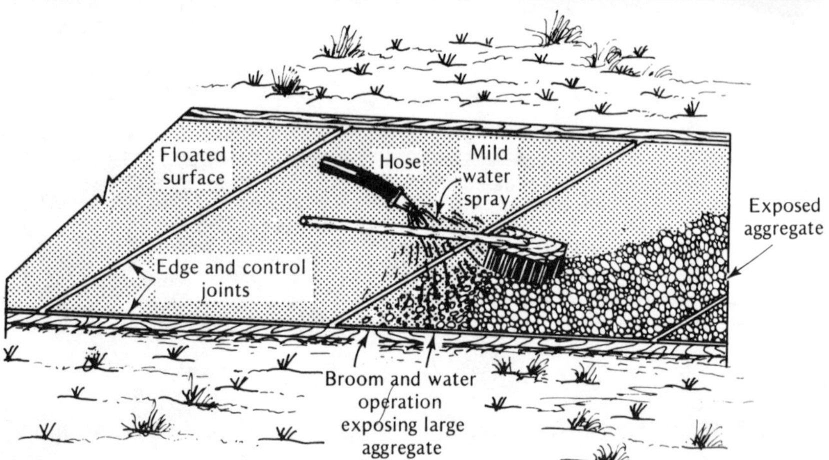

Figure 7–11 Exposed Aggregate Finish *(Photo Courtesy of Portland Cement Association)*

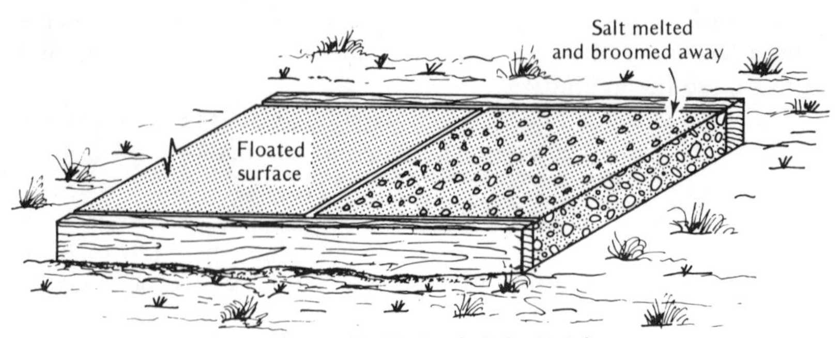

Figure 7–12 Rock-Salt Finish

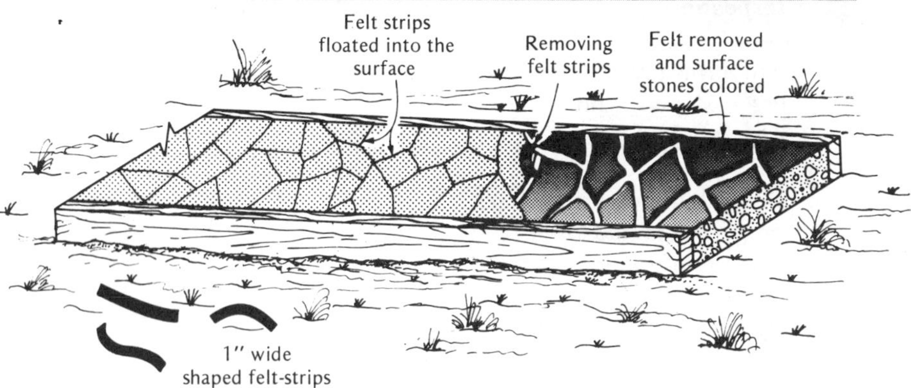

Figure 7–13 Making a Flagstone Effect *(Photo Courtesy of Portland Cement Association)*

Flagstone texture Strips of 15-lb felt cut 1 in. wide in desired patterns of curves and straight lines are pressed into the floated concrete surface (Figure 7–13). The surface is refloated; color dyes are added by sprinkling on the concrete and working the color into the concrete. Before the concrete is cured, the felt is carefully removed. (*Caution:* Do not chip the edges of the concrete when removing the strips.)

Curing Techniques

Where temperatures are mild and the relative humidity is other than very low, air drying is customarily used as a curing medium for sidewalks and driveways. This process, in the first 7 days, develops most of the compressive strength the concrete will obtain. Further curing during the next six months increases its strength.

Special precautions need to be taken to ensure that once curing is started, it continues even when adverse weather is imminent. Plastic sheet goods should be used to cover the fresh concrete during rain, freezing, or very hot weather. The plastic will prevent rain from ruining the concrete's surface texture. It will partially insulate the concrete during low temperatures and if a mat of straw is placed on top of the plastic, the finish is not disturbed. Finally, the nonporous plastic retains the moisture in the concrete that is needed for hydraulic action. (*Note:* See Chapter 1 for more on curing.)

INSPECTION OF SIDEWALK AND DRIVEWAY

Inspection	*Satisfactory/N.A.*	*Unsatisfactory*
Are forms properly located, level, sloped, and braced?	______	______
Are grades properly defined?	______	______
Has subgrade surface been properly tamped?	______	______
Has reinforcing material been cut and installed?	______	______
Have the proper type and quantity of concrete been ordered?	______	______
Was pouring done properly?	______	______
Were edging and center-joint control operations completed?	______	______
Was floating done properly?	______	______
Was final finishing completed for:		
a. Floating?	______	______
b. Troweling?	______	______

Inspection	*Satisfactory/N.A.*	*Unsatisfactory*
c. Brooming?	________	________
d. Special effects?	________	________
Was a curing technique properly applied?	________	________
Have forms been removed and earth backfilled?	________	________

QUESTIONS

1. What is edging?
2. Explain the use of 15-lb felt.
3. What is a subgrade?
4. What data about sidewalks and driveways are available from a plot plan?
5. What is the usual thickness of a sidewalk? Driveway?
6. What sidewalk or driveway data and detail are available from the elevation drawings?
7. Would it be usual to find a specification about sidewalks and driveways? What could it contain?
8. Name two materials that are used for forming sidewalks.
9. How frequently should braces be installed along a form?
10. Name and explain one method of subgrading the formed area of a sidewalk.
11. What condition indicates a need to reinforce a sidewalk?
12. What factor number do you use to calculate the number of yards of concrete needed for a 4-in.-thick sidewalk?
13. What is a control joint? How do you make it?
14. Does floating follow jointing?
15. Explain in general terms the following types of finishes.
    a. Exposed.
    b. Rock-salt texture.
    c. Flagstone texture.
16. What curing techniques should be used on sidewalks or driveways if very hot dry weather is anticipated?

Chapter 8

Concrete Steps

Dry mixture a mixture of concrete whose water content is severely restricted.

Precast concrete cast at an earlier time, cured, and made a part of a larger assembly.

Rise the number of inches or centimeters each step measures vertically from surface to surface.

Riser a board in a form equal in width to the rise in inches or centimeters; forming the rise.

Run the horizontal dimension of each tread in a staircase equal to the tread depth front to back.

Slump the amount of collapse plastic concrete has when not contained; its ability or inability to retain its shape.

Stringer a form member that contains rise and run dimensions and necessary anchorage for risers and bracing.

Tread that segment of a step that is walked upon.

OBJECTIVES—INTRODUCTION

When considered as a total job, the construction of a set of concrete steps is a very complex undertaking. It involves considerable planning, forming, and finishing techniques. Adding design or ornamental character complicates the tasks even more.

Planning includes determination of a proper rise and tread for the staircase. Then forming this requires laying out and cutting wood stringers and risers that support the freshly poured concrete. Floating and troweling are only two of the finishing techniques used. Therefore, a single objective can set forth all that must be done: *to be able to lay out, form, pour, and finish a set of concrete steps.*

An easier method and one that can be used where see-through stairs are desired involves the use of masonry stringers and concrete step treads or just treads. Both parts of this type of staircase are built separately. The risers are made from masonry blocks or bricks, and the precast treads are mortared in place. Planning is a prerequisite and forming is required, and this leads to the second objective of this chapter: *to be able to install a staircase using precast concrete step treads.*

THEORY OF STEP LAYOUT

Elevation Details

The elevation plan or a detailed sectional plan should show the exterior staircase. Whether it be a porch step or a full flight of steps leading to a basement from the outside, these plans detail much of the data needed by the builder. Study the partial plan shown in Figure 8–1 to locate the following items, then correlate their meaning as described below.

1. Total rise of staircase.
2. Total run of staircase.
3. Individual step rise.
4. Individual step tread depth (run).
5. Finished basement floor level.
6. Footing for landing.
7. Slab for landing.

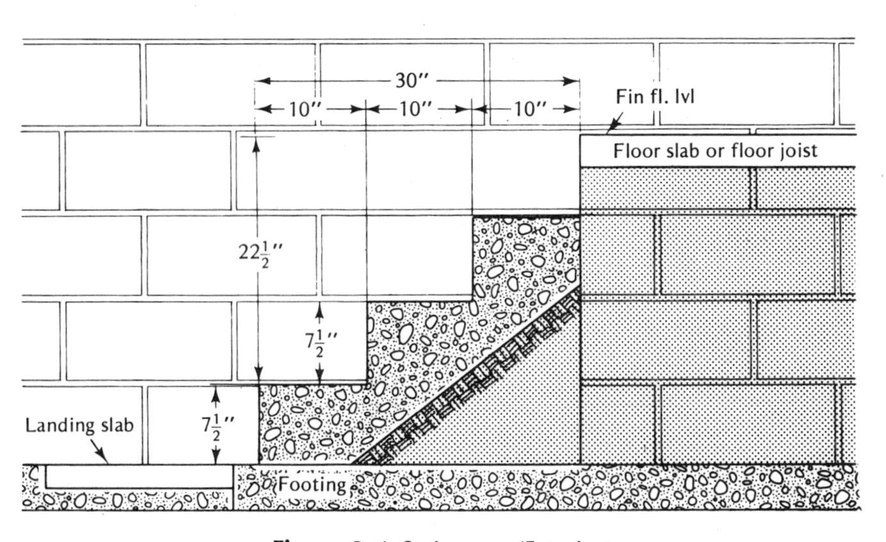

Figure 8–1 Staircases (Exterior)

The total rise of a staircase is the total distance between the lower landing and the top step or upper platform. This total number of feet and inches or meters in metric measurements will be instrumental in determining the average rise of each step.

The total run of a staircase is the horizontal distance from the leading edge of the first riser to the back edge of the top tread. This distance does not take into account any overhang of step tread. It does, however, consider the width of a tread at the top even if the tread is actually a finished floor. If the condition just illustrated exists, the staircase consists of one more riser than tread [e.g., 4 risers and 3 runs (treads), 5 risers and 4 runs].

The individual step (rise) is restricted in height by various building codes to a maximum of 7½ in. (16.5 cm) except for single-family dwellings, where an 8-in. (20 cm) rise per step is permitted. A rise consists of the distance from one tread surface to the next. It may be perpendicular, or beveled backward from the tread above to the tread below (Figure 8–2). If a beveled or recessed rise is planned, its form must be made accordingly.

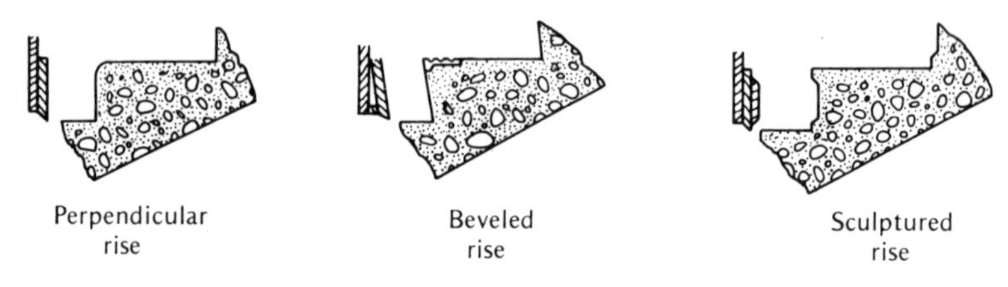

Figure 8–2 Riser Shapes

The individual step run or tread is the horizontal segment of a staircase. It must be a minimum of 10 in. (25.5 cm) from front to back with the exception that on single-family dwellings, a 9-in. tread is sometimes permitted.

The builder can always use distinct references, such as the finished basement floor level, footing, slab, or landing detail in the plan from which to base his measurements.

Rise vs. Run Criteria

The previous section identified the requirements for staircase rise as 7½ in. maximum and a 10-in. run as minimum (with exceptions noted). This combination results in a minimum of horizontal distance from first to last step. However, there may be several occasions where these values do not produce the most esthetic appearance.

Several alternatives are shown in Figure 8–3. Staircase A is a standard 7½-in. rise × 10-in. run; note that for staircase B, to rise to the same height one more rise is needed, and considerable more horizontal length is used. Staircase C with its 5-in. rise and 12-in. run takes even more horizontal distance. The question is: What values do these

Figure 8–3 Alternative Rises and Runs for Staircases

alternatives have? The answer lies in their purpose and artistic enrichment to the house and landscape. Clearly a smaller rise per step makes climbing easier for everyone, especially children, senior citizens, and handicapped persons. The added length develops a smooth graciousness to the approach of the home.

In contrast, a rise of less than 5½ in. and a tread width of more than 12 in. makes climbing uncomfortable. Too many steps makes one just as tired as too steep a rise, and too long a run may make the climber take two steps on one or more step treads during the ascent.

THEORY OF FORMING FOR ON-SITE POUR

The Form

Figure 8–4 shows a complete ready-to-pour form for a concrete staircase. Its elements consist of riser boards, stringers, and bracing. This form illustrates a bottom where the area under the steps is to be left open. If earth fill is used instead, the bottom boards need not be used. Rather the earth should be sloped and tamped to form the graded slope.

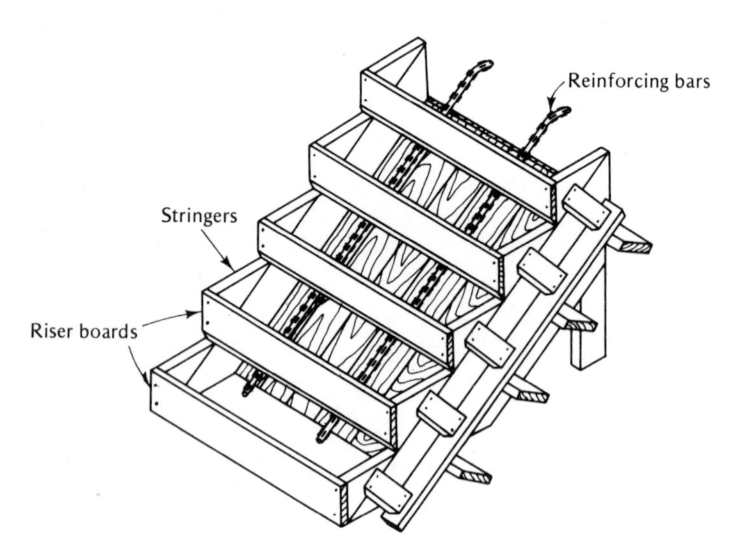

Figure 8–4 Complete Ready-to-Pour Concrete Staircase

The Riser

The riser shape shown results in a step rise that is perpendicular. To make a slant-under, the rise portion of the stringer needs to be cut under from several degrees up to 10 degrees. For a lip on the step run, an offset needs to be made into the riser board.

The forming method shown in Figure 8–4 allows masons free access to riser removal and finishing operations. By simply removing each riser board, the concrete behind it is exposed and can be floated and finished.

The Stringers

The stringers are usually built up, for two reasons: (1) 4 to 5 in. of concrete must be poured from the interior point of rise and run to earth or bottom form, and (2) nominal 2 × 12 stock may not be wide enough to make the stringer. The builtup method starts with a 2 × 6 intact or ripped to slab thickness (4 to 5 in.). Then rise and run wedges are laid out with framing square, cut, and nailed in place. Riser boards are prepared after the stringers are made, and all are nailed together.

The unit is then positioned and braced. During placement and while bracing is being added, many level checks should be made. The critical level checks are (1) placing the level on top of a riser, and (2) placing it on a run cut surface on the stringers.

THEORY OF POURING AND FINISHING CONCRETE STEPS

Subsoil Preparation

Since concrete steps often rest on the ground, backfill is used to reduce the quantity of concrete needed. Broken bricks, rocks, or broken concrete are used to fill the area beneath the steps. Then sand and fine gravel are shoveled on top and used to fill all holes and crevices. Water from a hose washes the sand well into these places, and tamping is used to firm the surface.

During this filling operation, grading for proper depth is continuously checked and corrections are made. Remember that the grade must be a minimum of 4 in. below the interior corner of riser and run.

Reinforcement

Where a set of steps are self-supporting, as shown in Figure 8–4, steel reinforcement bars are absolutely essential. A minimum of two bars is needed. More may be used if circumstances warrant. Placing bars several inches from the outer edges of the steps and on 16-in. centers throughout the width provides adequate tensile strength.

In contrast, where the staircase rests on packed earth fill, wire mesh common to that used in slabs and driveways should be cut and installed before the concrete is poured.

The actual steps are rarely reinforced. The weight force of persons walking on the treads is transferred to the bottom of the staircase where the steel is installed.

Placing Concrete

Concrete cannot be poured into the form. If the concrete were poured directly from a chute on the truck, the force could easily rupture the form. Therefore, it must be placed into the form carefully. Fill the bottom area first, then work up the stairs. Compact the mixture continuously as the concrete is being placed.

The mixture must be stiff. It must have a reduced water content. The stiff mixture has very little slump, and this character is necessary to prevent overspill of concrete from lower step treads. Consider what would happen if a standard or wet mixture were used. As more concrete was placed in the third, fourth, or fifth steps, it would force lower-placed concrete to ooze out over the lower steps.

Finishing, Including Riser Removal

Floating and edging operations are used to finish the stair treads (runs) and risers. First, when the concrete has lost its sheen and is stiff enough to support weight without creating a depression over ¼ in. deep, the tread surfaces should be floated. If rounding of the lip is needed, the edge tool should be used to round the edge concurrently with the floating operation.

Next, the top riser is removed and the concrete behind it is floated to a finish. The next lower riser is removed and the concrete is floated and so on, until the bottommost rise is finished.

The staircase should be covered with plastic to aid in curing the concrete. The plastic retains the moisture in the concrete and allows hydration to continue.

THEORY OF POURING AND INSTALLING PRECAST CONCRETE STEP TREADS

There may be one or more reasons why the poured method of building steps is too difficult to construct. For these reasons an alternative method is presented.

Building Risers from Other Masonry Materials

Figure 8–5 shows that bricks or blocks can easily be used to create the risers and run support surfaces needed. First a footing is needed to support the bricks or blocks. Then the bricks or blocks are laid in courses, stepping back for each new riser. When both parts are complete and dry, the treads should be mortared in place.

Step-Tread Forming and Pouring

Figure 8–5 also shows the form that needs to be made from nominal-2-in. stock according to the dimensions of the step's tread width, length, and thickness. A full 2 in. should be the minimum thickness, so a 2 × 6 ripped in 2-in. pieces develops the form's height (tread thickness). These pieces are then cut to the necessary lengths and nailed together to form a rectangle. A plywood bottom may be cut and nailed to the form to square the form and complete it.

If the bottom is installed, a very dry mixture of water, cement, fine aggregate, and small pea aggregate well mixed is placed in the form along with two steel rods for reinforcement. Tamping methods should compact the mixture into a precise shape of the form. Then it can be turned over and lifted carefully. The concrete step is left on the surface to dry, and the form is readied for immediate use. This method allows making all step treads during a single operation.

If the form is made without the bottom installed, the form should still be filled with the dry mixture just illustrated. Then when the form is full, its surface should be floated and the form carefully lifted.

The stiff mixture will hold its shape and curing does take place. In 7 days the treads are ready to move to the steps, where they are mortared in place.

Installing the Precast Treads

A standard batch of mortar should be mixed. Then a ½-in.-thick bed of it should be laid on the brick or block where the precast tread will be placed. The tread is placed on the mortar and leveled with a spirit level. All excess mortar should be removed with a trowel so that a clean, flush joint is made.

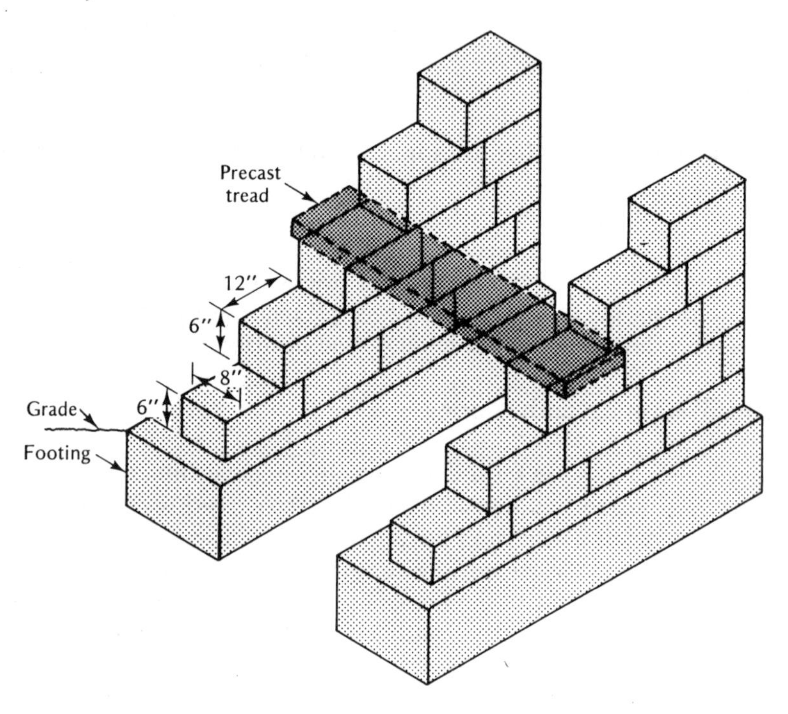

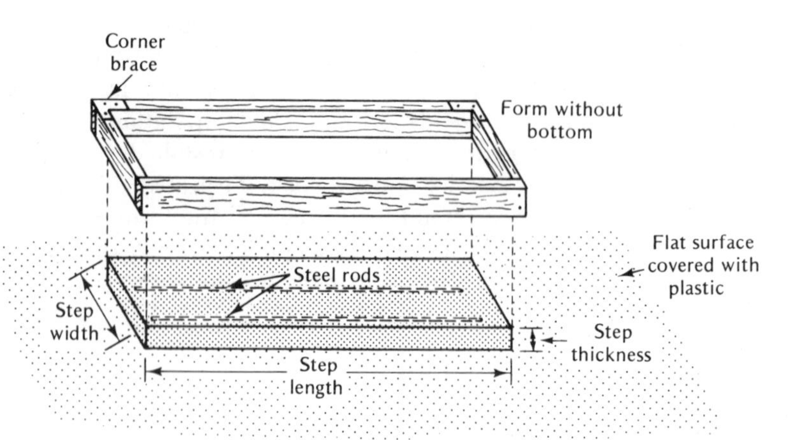

Figure 8–5 Steps Using Precast Treads

INSPECTION OF CONCRETE STEP CONSTRUCTION

Inspection	*Satisfactory/N.A.*	*Unsatisfactory*
Are rise and run dimensions according to acceptable standards?	________	________
Are forms properly prepared, aligned, level, and perpendicular and braced?	________	________
Is backfill properly compacted and graded?	________	________
Have reinforcement bars been positioned properly?	________	________
Has a dry (stiff) mixture of concrete been prepared?	________	________
Was compacting performed while placing the concrete?	________	________
Was floating of runs followed by floating of rises done properly?	________	________
Was a proper curing method employed?	________	________
Were forms removed and exposed surfaces grouted?	________	________
If precast step treads were used:		
a. Were they reinforced?	________	________
b. Were they cured properly?	________	________
c. Were they mortared in place properly?	________	________

QUESTIONS

1. Explain the difference between rise and run on a staircase.
2. What is a stringer?
3. Is there a common usage for run and tread?
4. Which plan provides individual rise and total rise data?

5. Is the total run of a staircase a vertical or horizontal dimension?
6. What are industry standards for maximum rise and minimum run in inches? In centimeters?
7. Can the rise be more than 7½ in. for residential stairs?
8. What advantage would making the rise less than 7½ in. provide?
9. What function does the riser perform?
10. When pouring a set of concrete steps, why should the mixture of concrete be dry?
11. Which riser is floated first, the top or the bottom?
12. What is the advantage of using a precast step tread?
13. In constructing a stair with precast treads, what are some materials that the mason can use to develop the rises and runs needed?
14. How are precast treads installed?

Section II

Bricks, Blocks, and Stones

From the earliest times to the present, the skill, sometimes called the "art," of bricklaying has been in continuous use. Masons the world over have used bricks made from a wide variety of materials to build structures.

During earlier times bricks were molded from clay or adobe and mixed with other substances to give them the desired color, consistency, and weight. These molded bricks were baked in ovens or left in direct sunlight during intensely hot periods of the day. They were then mortared into place.

Today bricks are molded from clay and other minerals and substances, colored, and baked in ovens. They, too, are mortared into place. So one can assume, correctly, that a product that has stood the test of time so well should continue to be made and used today.

As in Section I, the mason uses a set of blueprints as his guide not only for bricklaying but also when laying block and rubble stone. The principles of laying masonry and natural materials are included in a set of plans. Some details are given in foundation and floor plans; other details are given in elevation and sectional drawings. Some of the symbols used for these masonry materials are shown in Figure A–1. In addition, some details are given in specifications because they are not easily shown in blueprints.

The various chapters in this section contain data related to understanding such blueprints and specifications. From the understanding developed, the reader will become familiar with the principles of brick, block, and rubble-stone masonry.

In addition to gaining a knowledge of the principles defined above, a vocabulary of terms is defined and used in illustrations. These include names associated with methods of laying brick, such as the

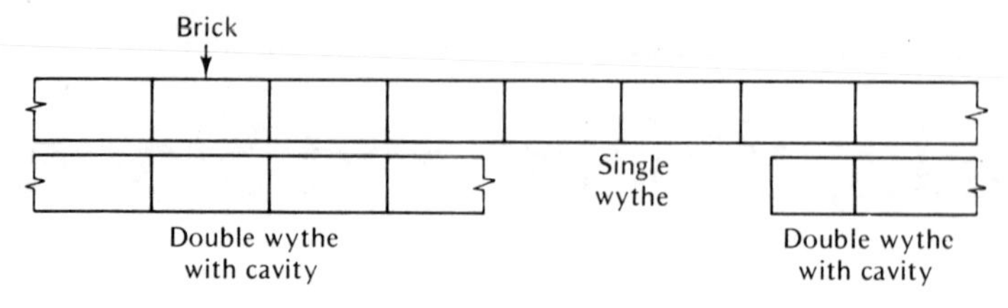

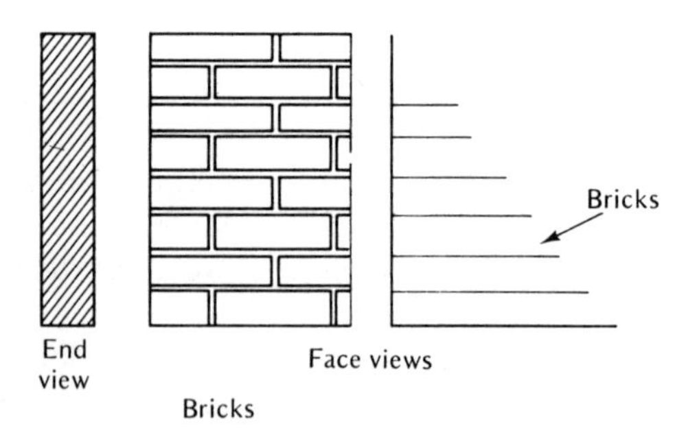

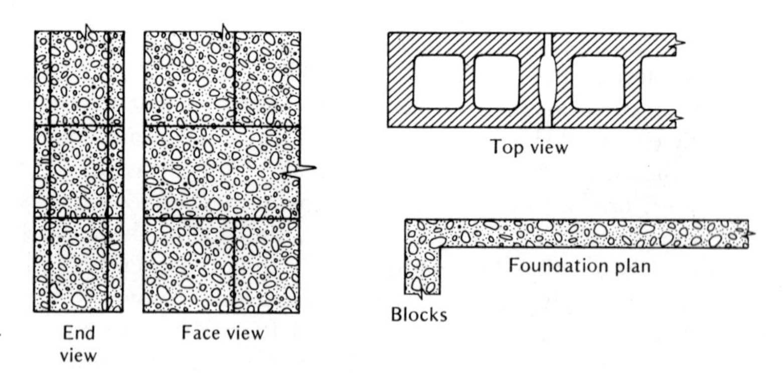

Figure A-1

stretcher, the *header*, and the *soldier* course (note in Figure A–2 that the bricks standing on end resemble a line of soldiers).

Each chapter would be incomplete if instruction were not given on the architectural beauty that can be obtained from the masonry products being used. Many patterns of laying brick and block are shown. Two patterns of laying rubble stone are shown.

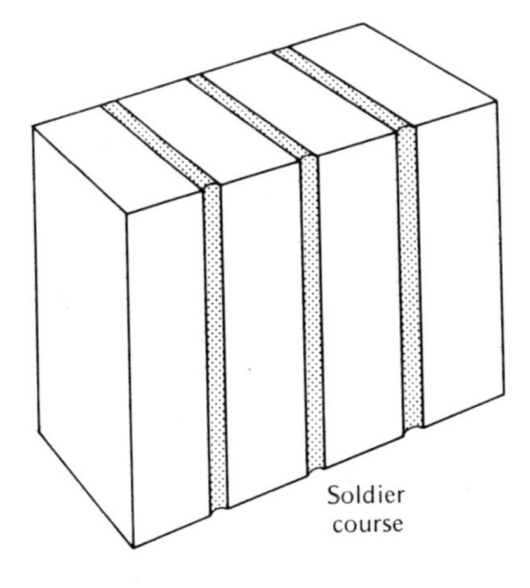

Figure A-2

Techniques of laying these masonry materials are shown and explained so that the reader is able to grasp the various movements used by the mason as he lays his brick, block, or stone. Such instruction includes the laying of mortar beds, buttering the ends of masonry units, laying mortarless pavers, and many other details.

Finally, each chapter ends with a quality inspection checklist. The list is structured so that a logical progression of work ensues, and therefore the novice and apprentice can expect to see a logical progression by studying the list.

Chapter 9

Brick Walls

Bonded the union, with mortar, of two or more bricks, courses of bricks, or wythes.
Cornice wood materials used to make a finish on a building between a wall's top edge and a roof's lower edge.
Frieze one board in a cornice, usually just above the last course of brick.
Header a brick installed at right angles to the run of the wall.
Lintel a steel angle iron placed over the door and window frame.
Mortar a mixture of cement, lime, and sand used as a bonding agent in bricklaying.
Rowlock bricks set on their long side; normally used in windowsill construction.
Sill the lower portion of a door or window.
Stretcher bricks laid end to end with the long side facing outward.
Ties metal or headers used to secure wythes and provide lateral strength.
Veneering the installation of a single wythe of brick on a building and using wire ties to that building.
Wythe a vertical stack of bricks one thickness wide (e.g., a veneer course).

OBJECTIVES—INTRODUCTION

Building a brick wall or installing a single width of brick on the face of a frame building, *veneering*, is a complex task. The bricks are small and easily handled, and mortar is easily proportioned and mixed to proper consistency; yet using these materials requires practice to obtain the needed skills. Also, much planning is needed so that the correct conditions exist.

These inherent needs lead to the two objectives for this chapter, which are: *to understand the principles of brick wall construction,* and *to be able to identify the practical tasks and needs in laying a brick wall.* To relate understanding more directly to actual building, segments of blueprints and specifications are used. Then practical tasks used in constructing walls from brick are examined, and the chapter ends with the inspection checklist.

READING AND UNDERSTANDING BLUEPRINTS AND SPECIFICATIONS FOR SOLID BRICK WALLS AND VENEER BRICK WALLS

Brick veneering the outside of a house, for instance, requires one set of elevation detail drawings, and a solid brick wall requires another. Either or both may additionally need specifications to set standards or call out requirements that cannot be shown in plans.

The overall elevation plan for a brick house may show very little detail except closely spaced horizontal lines plus the notation "brick," because the plan shows a full length and height of a wall in several square inches. This, of course, makes adding detail very difficult. Therefore, the architect renders detailed plans of the critical areas to aid the brick mason. These are examined in depth.

Brick Veneering

Several different approaches may be used to establish a base (foundation) for brick veneering. In each the brick is laid on solid concrete or a concrete block surface. For the first example, study Figure 9–1. The elevation drawing is one of a poured slab with a 5-in. inset that is used as a ledge to lay the bricks upon. Locate and observe the following details that are given to the mason: (1) brick veneering is to be installed; (2) galvanized wall ties are used to tie the brick to the stud wall; (3) there is to be a 1-in. air space between brick and sheathing; (4) the wall consists of framing and sheathing; (5) weep holes are made into the mortar each 4 ft 8 in. o.c., and either 30-lb felt or hemp rope is used to guide condensation within the wall to the outside. With this information, the mason proceeds to lay the brick. Figure 9–1b illustrates the translation of elevation details as the mason understands them. Notice that the wall ties are installed and that the weep holes are evident. The bricks are laid in horizontal "stretcher" course formation with all vertical joints perpendicular and plumb.

A somewhat different arrangement is illustrated in Figure 9–2. The brick is laid directly on the footing and, as is shown, several courses

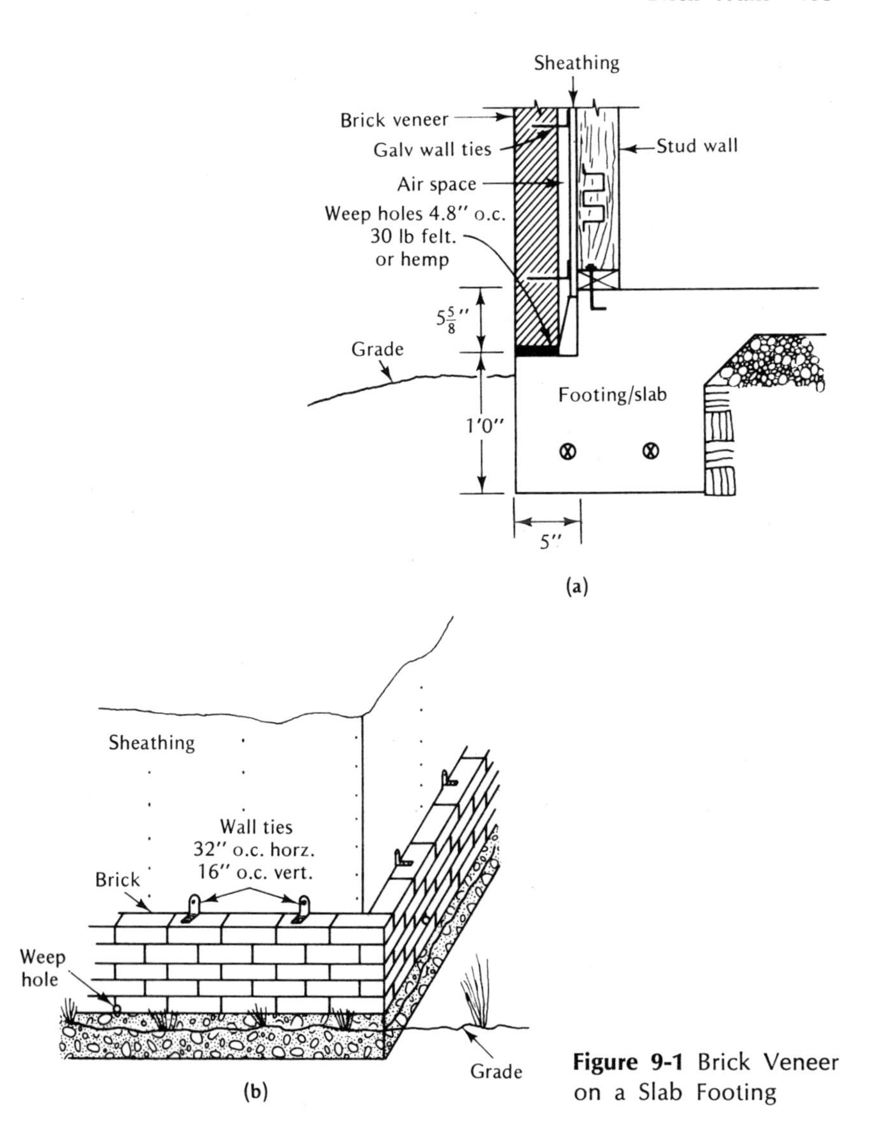

Figure 9-1 Brick Veneer on a Slab Footing

of brick are below finished grade level. The elevation detail shows a block foundation, backfilled with dirt, then a poured slab floor. It shows a framed wall with sheathing.

When compared with the detail plan used in Figure 9–1, this one has exactly the same data. Galvanized ties are called for; so is the 1-in. air space, and weep-hole detail is added. However, it appears as if the weep hole is in a different place. Actually, it is several courses

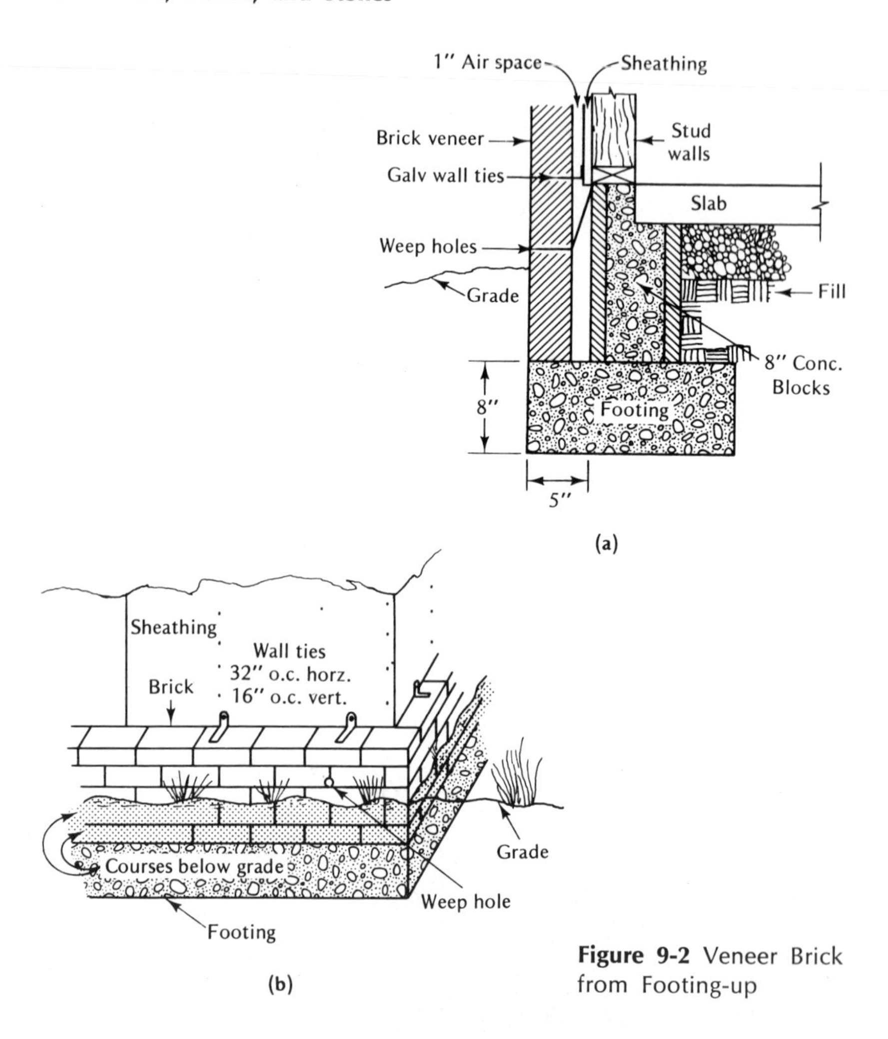

Figure 9-2 Veneer Brick from Footing-up

above the footing; but it must be several inches above finished grade level, and in light of this it is properly located. Figure 9–2b shows what the mason translates from the plan.

A third type of detail drawing is shown in Figure 9–3. This one is customarily used where basements are installed below ground and exposed walls are to be brick-veneered. Block walls are raised by masons with a solid 4-in. cap block laid last. The sill, joist, sheathing, and wall are kept back 5 in. from the outside edge of the block to allow for the brick and a 1-in. air space.

Details paralleling those shown in Figures 9–1 and 9–2 are provided. In this circumstance the weep holes can and should be made

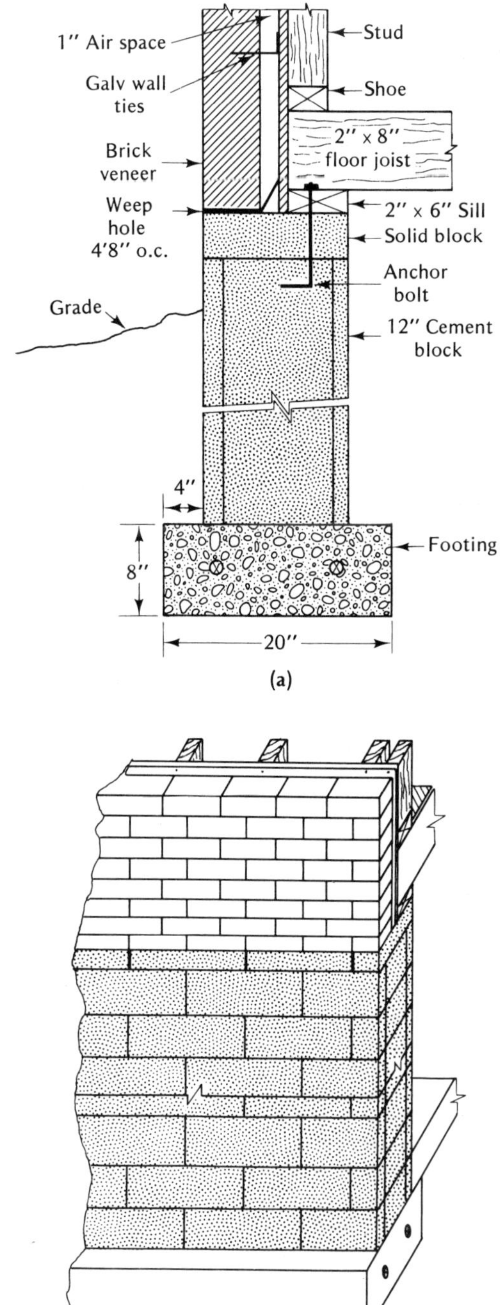

Figure 9-3 Brick veneer on a Foundation Wall

in the base mortar of the first course. Figure 9–3b shows the wall as the mason translated the details.

The next detailed section of an elevation drawing is the windowsill. When brick veneer is used on the wall, the sill is most often made from the same brick. These are laid on edge in *rowlock* fashion rather than flat. They are also installed on a slant in the direction of the rainfall, as Figure 9–4 illustrates. Study the elevation details closely and note the following. The brick is veneered below the window and up the walls on both sides of the window; the sill brick is sloped and inserted under the window. This is very important to a sound, waterproof job. The remainder of the details are the same as those in

Figure 9–4 Brick Veneer on Windowsill

Window
Window sill
Sloped for drainage
Stud
Sheathing
1" Air space
Wall tie
Brick veneer

(a)

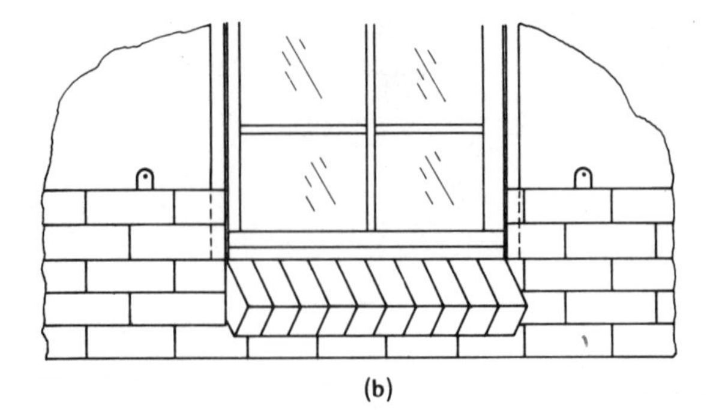

(b)

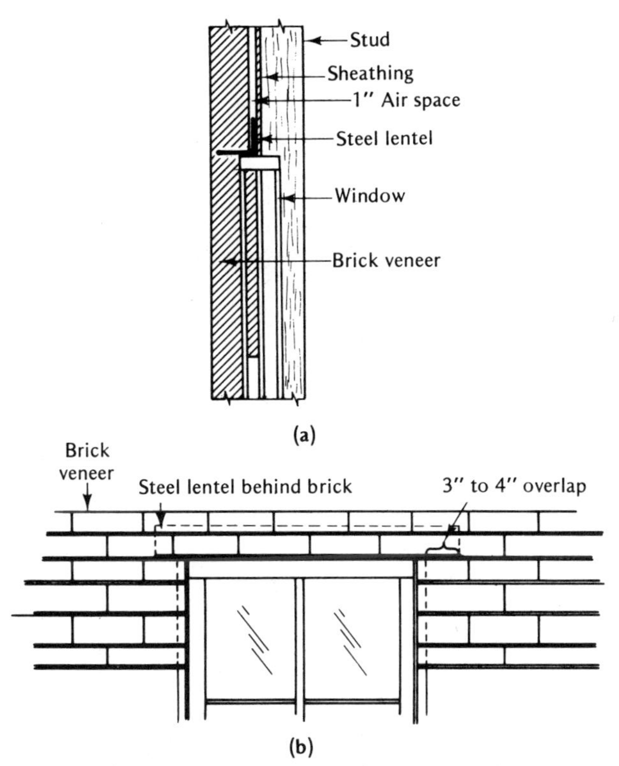

Figure 9–5 Brick Veneer over a Window (Same for over a Door)

Figures 9–1, 9–2, and 9–3; they are the 1-in. air space, wall ties, sheathing, and studs.

Bricks that are laid above the window (and door frame as well) cannot support themselves, and these units cannot support the brick and mortar weight and force. Therefore, a lintel made from a steel angle iron is used as a base. The lintel is set onto the course of brick even with the window or door head. It must overlap the bricks by 3 to 4 in. on both sides of the window. Figure 9–5 shows these details. Notice that the lintel is placed as stated earlier and that the bricks are continued across the window without interruption of the stretcher course design. (Mortar is laid directly on the lintel.)

The brick veneer must unite with the house cornice regardless of the type of cornice. It must be installed so that the frieze board on the cornice overlaps the top course of brick by not less than ½ in. Figure 9–6 illustrates the elevation details of a boxed, overhanging cornice, brick veneer, stud, and rafter. Study the position of the frieze

Figure 9–6 Top Course of Brick and Cornice

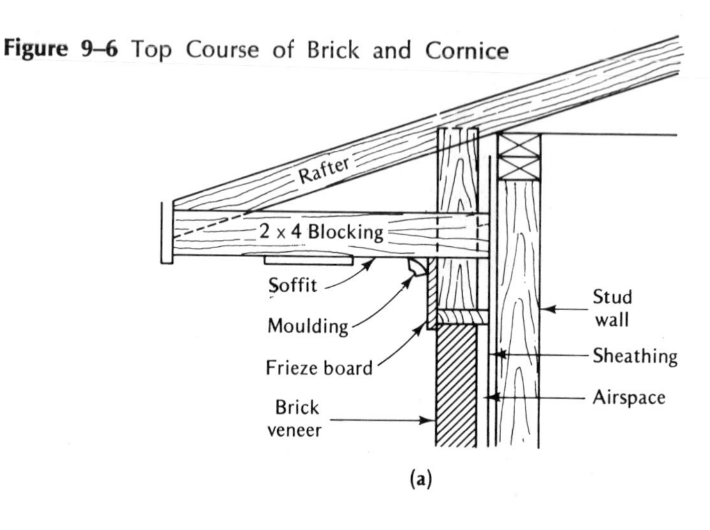

(a)

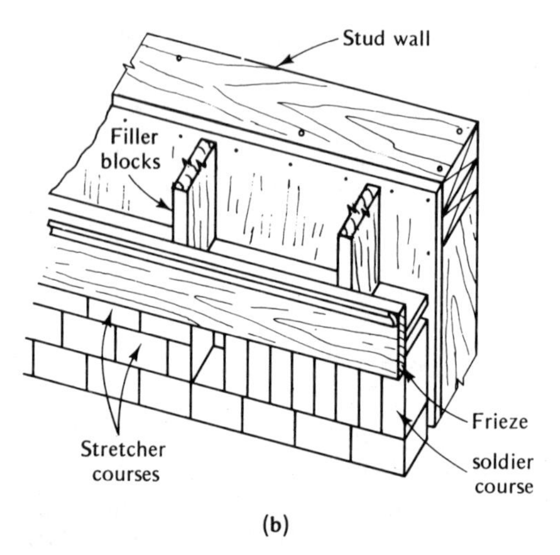

(b)

board and note its position overlapping the top course of brick. Carpenters install blocking, usually 2 × 4s, to provide a sound nailing surface for the frieze board. Either the *stretcher* or the *soldier* course is used as the top course, as Figure 9–6b shows. The requirement for the course to extend above the lower edge of the frieze board by at least ½ in. must be met regardless of the design being used.

Most of the detail shown in the drawings can often be found as shown; however, it need not all be there. For instance, the sill details could easily be included in the specifications as: "Make sills from the same brick as used for veneer." The top course design feature could be stated as: "Install a soldier course as the top course." Lintel detail could also be given in the specifications as: "Use ¼-in. lintels over windows or doors and cut them 6 in. (or 8 in.) longer than the window width." Instructions for weep holes, mortar, type of brick pattern, and type, size, and gauge of wall ties could also be given in the specifications.

Solid Brick Wall Construction

Modern solid brick walls, even walls with cavities between wythes, use a poured footing made from concrete. In one operation a unified reinforced structurally sound base is established. The footing is made wide enough for a double thickness of brick plus the required overlap on either side of the brick. Its depth is usually 8 in., but on occasion it is thicker, and this thickness makes it sufficient to support the load of the wall. Reinforcing rods must be installed to provide needed tensile strength to aid in prevention of stresses within the brick wall that could result in joint cracking or joint separation.

Laying a solid brick wall involves considerably more study and planning during its design and construction. No framed wall can be used to support it; therefore, it must be self-supporting. Two wythes (rows) of brick are laid side by side; neither is sufficient in strength to withstand lateral forces. So unions must be incorporated to develop needed strength. The four common methods used to develop the strength are shown in Figure 9–7. The header course, in a single, double, or continuous arrangement, ties both wythes together by bridging from one to the other and by being held in place with mortar. Headers are used in numerous places, but as ties they are customarily installed every *fifth* course. To be sure that sufficient headers are used, use the following standard: "4 percent or more of the exposed surface should be composed of headers; or the distance between adjacent full-length headers (courses) should not exceed 24 in. vertically."[1] The 4 percent allows design and implementation of a variety of patterns where headers can be incorporated for pleasing appearance.

Metal ties, such as wire lath, shaped galvanized, and bent rods, can be used to unite the wythes. They must be installed similarly to headers, such as every fifth vertical course and continuous or spaced every 24 in. horizontally.

Details about ties may be included in specifications only, but they could also be given in elevation detail drawings. If headers are

---

[1] Standard No. 2412, *Uniform Building Code*, 1973 ed.

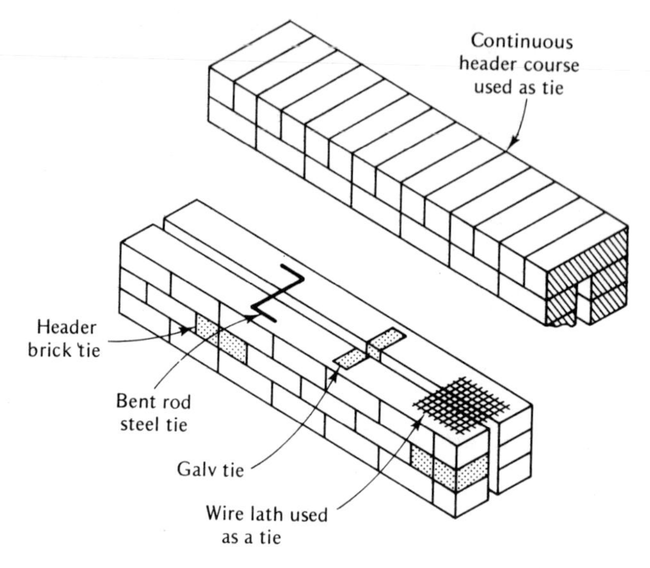

Figure 9–7 Types of Wall Ties for Brick Walls

used, a full design section could be rendered so that no misunderstanding of the pattern results.

Elevation Details for Solid or Cavity Brick Walls

Beginning construction of a wall requires the use of foundation details such as those given in Figures 9–8 and 9–9. Figure 9–8 provides data on beginning construction of a cavity wall and Figure 9–9 provides data on beginning a solid brick wall.

Close examination of the details given in Figure 9–8 points to: a 4-in. overlap of footing either side of the brick, a header course as a first course embedded in mortar, wall ties of metal, and a 1-in. air space (cavity). In this arrangement both wythes are united and reinforced by the metal ties. The drawing to the right (Figure 9–8b) illustrates the mason's interpretation of the detail plan.

The major difference between the wall shown in Figure 9–9 and the one shown in Figure 9–8 is that there is no cavity. Bricks are laid and mortared so that their inner vertical surfaces are bonded. Notice that details indicate that the first and top courses are both header courses. There will, of course, be intermediate headers throughout the wall.

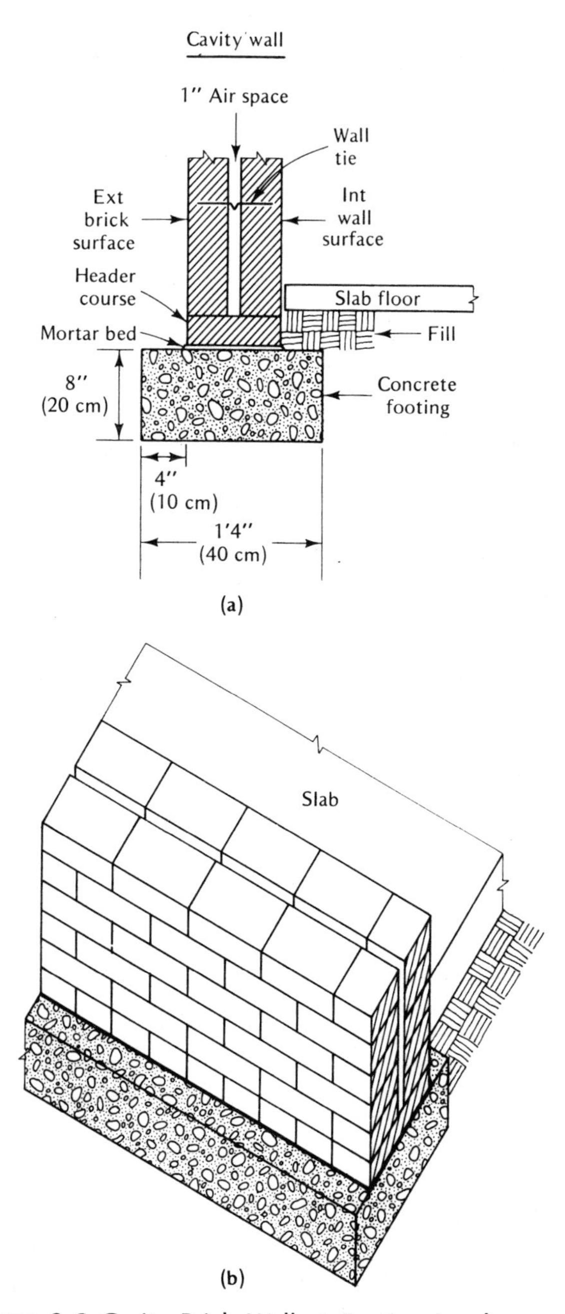

Figure 9–8 Cavity Brick Wall at Footing Level

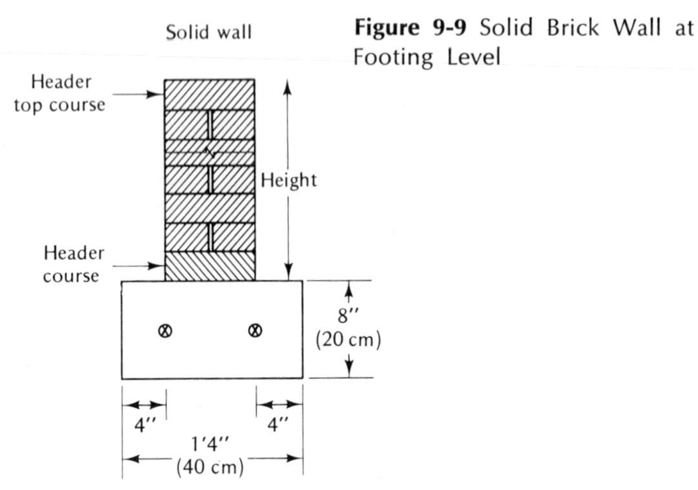

Figure 9-9 Solid Brick Wall at Footing Level

Sectional Details for Brick Walls Around Doors and Windows

The data presented here are equally applicable to the cavity and solid types of brick wall construction. Therefore, only one type of illustration is present. Figure 9–10 provides elevation data for door frame, door sill, and door head. The details for door head lintel are also applicable to window head and lintel construction.

Locate the data given in the lower portion of Figure 9–10a: footing, header course, door sill, door frame consisting of door brick trim, door stop, and jamb. These data have meaning for two reasons: (1) the sill height is given as *flush* with the slab floor and flush with the inner wall surface; and (2) the door frame is used as a guide by brick masons. This means that the frame is placed where the plan calls it out and is braced plumb and true. The masons then cut and mortar their bricks to butt against the trim.

It is possible, however, to brick a wall and allow openings. If this method is employed, anchoring devices such as bolts or filler wood blocks are installed. The door jamb is then fastened to these. Either method is appropriate, but the second one requires extreme accuracy, whereas the first method simplifies the bricklaying task.

Figure 9–10c shows a section of wall and door frame and sill as it looks when completed. Notice how the sill slopes away from the edge of the slab. Also notice that the brick butts against the wood brick trim, yet there is space between the jamb and brick for a wood filler block if necessary.

The elevation drawing of Figure 9–10 also details the head and lintel needs as well as final header course, anchor bolt, and wood plate. Since this detail plan is for a solid brick or cavity-type wall, two lintels are needed. Each angle supports all bricks on its respective side of wall center. Therefore, the vertical part of the angle must be in the

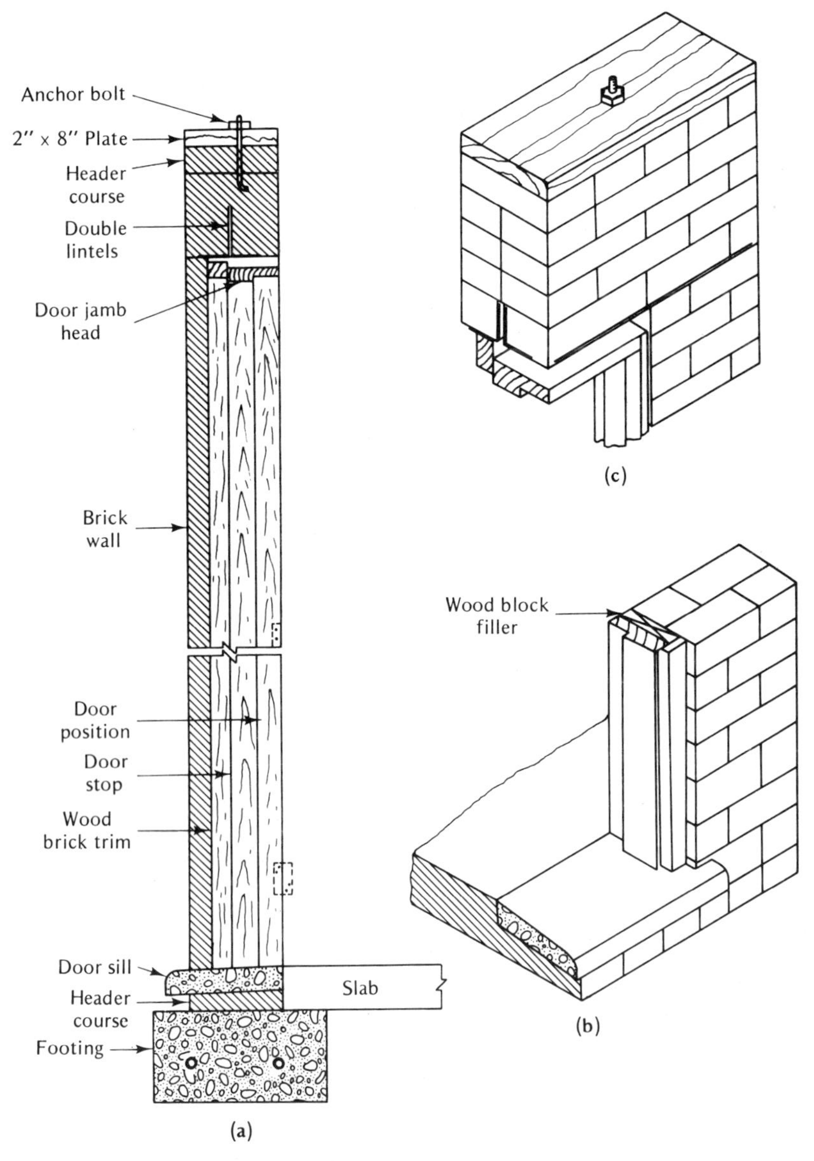

Figure 9–10 Elevation Details for Door and Window Brick Work

center of the wall. The lintel needs to be made from two 3 in. × 3 in. × ¼ in. angles for any opening up to 5 ft across. Openings wider than 5 ft and up to 8 ft need two lintels 3½ in. × 3½ in. × ¼ in. (*Note:* For 12-in.-thick walls, use lintels ½ in. wider than specified above.)

The lintel carries the weight and force of all bricks installed over it. This is correct, because the door or window frame cannot do the job. It is important to note, though, that the normal installation of the door's wood brick trim on the head extends above the door head jamb, thereby creating a space between head and lintel. This space does not need to be filled with blocking material, since door trim on the inner surface of the doorjamb and brick wall will cover the space. Figure 9–10 shows the details as they were translated; every detail is included.

Capping the wall in preparation for ceiling joists and rafters requires the need for a wood plate, as Figure 9–10a shows. The details for the final course of brick and plate indicate that the final course is to be a continuous header course. During its installation anchor bolts need to be installed. The wooden 2 × 8 plate is installed after the mortar has set to prevent any interruption to hardening or curing.

The final major area of bricklaying detail is illustrated in Figure 9–11 and concerns the windowsill. Generally, the height of two courses of brick is used to allow sufficient space for windowsill construction. The general rule is that brick is installed in *rowlock* fashion, and the top surfaces tilt in the direction of rainfall. Since the rowlock course extends out from the wall several inches, a cavity is created along the

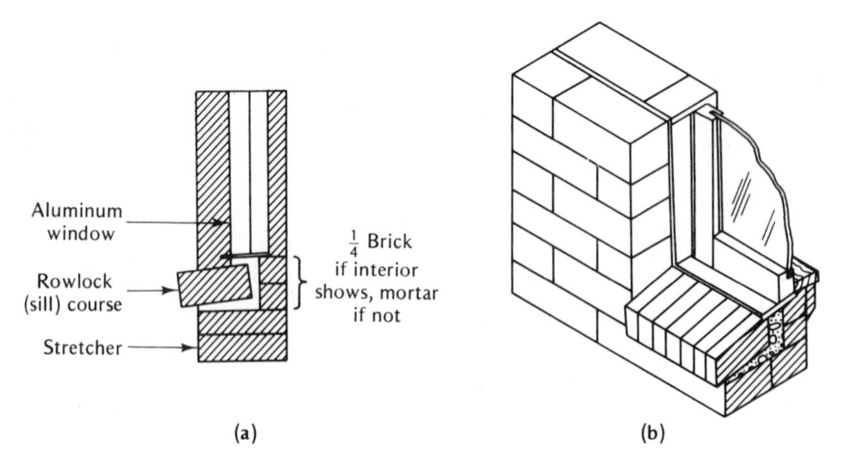

Figure 9–11 Elevation Data—Windowsill Area

inner wall. This cavity can be treated in two ways. The method shown in Figure 9–11a indicates that a 1/4 wide brick shall be installed two courses high. The windowsill is installed over the top course of brick. Another method, which can be used if the inner surface of the area is to be covered with another building material, is to fill the cavity with mortar. When doing this a formboard is held in place against the inner brick wall; then mortar is placed into the cavity. The inner sill and apron (molding below the sill) cover the mortar fill. Figure 9–11b shows the completed job.

PRACTICAL KNOWLEDGE NEEDED FOR BRICKLAYING

There are a variety of knowledges that every brick mason needs to successfully lay bricks or to understand the principles of bricklaying. Several of the most vital are examined in this section. They are mortar mixtures, brick-course layout, estimating brick quantities, bricklaying patterns, and joint-striking techniques.

Mortar Mixtures for Laying Bricks

Mortar is the binding agent that causes bricks to remain intact in place. It should be easily worked, yet stiff enough to hold up the brick and allow work to continue. For ordinary service or general use, a mixture of 1 part masonry cement (prepackaged) with 2 1/4 to 3 parts of damp loose mortar sand by volume is mixed with clear tap water. If portland cement is used in place of masonry cement, 1/2 to 1 1/4 parts of hydrated lime need to be added to 1 part of cement, and the sand volume is increased to 4 1/2 to 6 parts.

The type of mortar just described is named *Type M;* there are others, named *Type S*, *Type N*, and *Type O*. Each uses the same mixtures by volume, but their relative strengths are less. Type S is usable for general purposes and where lateral forces are experienced. Type N is used above ground level on exterior walls where it is exposed to extreme temperatures. Type O is used for load-bearing walls in relatively dry and moderately temperate environments.

The standard method used in mixing mortar is to measure out a volume of sand and mortar mix that can be used within a 2-hr span of time. The elements must be thoroughly mixed, creating a uniformly gray mixture. Water at 50 to 70°F is added in moderate quantities until a workable mixture results. This preparation can be mixed on a mortarboard if only several (25 to 50) bricks are to be laid. It could be prepared in a wheelbarrow if 50 to 150 bricks are to be laid, and it should be made in a mortar box if a greater quantity of brick is to be laid.

The mortar should be used when ready and well before it begins to set. Therefore, it may be necessary to determine the course layout before mixing the mortar or at least while the mortar is being prepared.

Brick-Course Layout

Several authorities on bricklaying state that the better bricklaying installations are made with ⅜-in.-thick mortar mixtures between courses of brick. However, it is frequently more prudent to use slightly thicker mortar between courses. See Tables 9–1, 9–2, and 9–3, which provide a variety of joint sizes. Table 9–1 gives data for a ⅜-in. joint, Table 9–2 for a ½-in. joint, and Table 9–3 a ⅝-in. joint.

How does one make use of these tables in laying out a brick installation? Elevation measurements as shown in Figure 9–12 provide the basis of the layout. From these measurements and the tables, a story pole is laid out and used during the installation.

The first overall measurement indicates that bricks are to be laid to a height of 8 ft 3 in. or (250 cm). According to Table 9–1, this height requires 37 courses of brick with ⅜-in. mortar between each course for a total height of 8 ft 1⅜ in. This means that the top of the final course is 1⅝ in. below the needed height. If a ½-in. joint is used, Table 9–2 shows that 36 courses are used and finish at exactly 8 ft 3 in. Finally, if a ⅝-in. joint is used, 34 courses bring the top course to 8 ft 1¾ in. high, or 1¼ in. lower than needed.

Assume that ½-in. joints are to be used. The next consideration is: How does the brick lay out for windowsill and head? From foundation to windowsill is a distance of 60 in. (150 cm). According to Table 9–2, 22 courses of brick are needed to reach windowsill height. Two courses below the actual window are omitted so that a rowlock can be installed later. The window is 31 in. (77.5 cm) high, and 11 courses bring the total course height to window head height. Finally, three more courses are used above window height to complete the wall. In this particular examination, the story pole is laid out with no variations. This is not to say that such a perfect condition always exists. From time to time, slight variations in joint thicknesses need to be made. Critical points usually dictate the adjustments to joint thickness that need to be made, as just illustrated.

TABLE 9–1 TOTAL HEIGHT OF COURSES USING 2¼-IN. BRICK AND ⅜-IN. MORTAR JOINTS

Courses	Height	Courses	Height	Courses	Height	Courses	Height	Courses	Height
1	0′ 2⅝″	21	4′ 7⅛″	41	8′ 11⅝″	61	13′ 4⅛″	81	17′ 8⅝″
2	0′ 5¼″	22	4′ 9¾″	42	9′ 2¼″	62	13′ 6¾″	82	17′ 11¼″
3	0′ 7⅞″	23	5′ 0⅜″	43	9′ 4⅞″	63	13′ 9⅜″	83	18′ 1⅞″
4	0′ 10½″	24	5′ 3″	44	9′ 7½″	64	14′ 0″	84	18′ 4½″
5	1′ 1⅛″	25	5′ 5⅝″	45	9′ 10⅛″	65	14′ 2⅝″	85	18′ 7⅛″
6	1′ 3¾″	26	5′ 8¼″	46	10′ 0¾″	66	14′ 5¼″	86	18′ 9¾″
7	1′ 6⅜″	27	5′ 10⅞″	47	10′ 3⅜″	67	14′ 7⅞″	87	19′ 0⅜″
8	1′ 9″	28	6′ 1½″	48	10′ 6″	68	14′ 10½″	88	19′ 3″
9	1′ 11⅝″	29	6′ 4⅛″	49	10′ 8⅝″	69	15′ 11⅛″	89	19′ 5⅝″
10	2′ 2¼″	30	6′ 6¾″	50	10′ 11¼″	70	15′ 3¾″	90	19′ 8¼″
11	2′ 4⅞″	31	6′ 9⅜″	51	11′ 1⅞″	71	15′ 6⅜″	91	19′ 10⅞″
12	2′ 7½″	32	7′ 0″	52	11′ 4½″	72	15′ 9″	92	20′ 1½″
13	2′ 10⅛″	33	7′ 2⅝″	53	11′ 7⅛″	73	15′ 11⅝″	93	20′ 4⅛″
14	3′ 0¾″	34	7′ 5¼″	54	11′ 9¾″	74	16′ 2¼″	94	20′ 6¾″
15	3′ 3⅜″	35	7′ 7⅞″	55	12′ 0⅜″	75	16′ 4⅞″	95	20′ 9⅜″
16	3′ 6″	36	7′ 10½″	56	12′ 3″	76	16′ 7½″	96	21′ 0″
17	3′ 8⅝″	37	8′ 1⅛″	57	12′ 5⅝″	77	16′ 10⅛″	97	21′ 2⅝″
18	3′ 11¼″	38	8′ 3¾″	58	12′ 8¼″	78	17′ 0¾″	98	21′ 5¼″
19	4′ 1⅞″	39	8′ 6⅜″	59	12′ 10⅞″	79	17′ 3⅜″	99	21′ 7⅞″
20	4′ 4½″	40	8′ 9″	60	13′ 1½″	80	17′ 6″	100	21′ 10½″

TABLE 9–2 TOTAL HEIGHT OF COURSES USING 2¼-IN. BRICK AND ½-IN. MORTAR JOINTS

Courses	Height	Courses	Height	Courses	Height	Courses	Height	Courses	Height
1	0′ 2¾″	21	4′ 9¾″	41	9′ 4¾″	61	13′ 11¾″	81	18′ 6¾″
2	0′ 5½″	22	5′ 0½″	42	9′ 7½″	62	14′ 2½″	82	18′ 9½″
3	0′ 8¼″	23	5′ 3¼″	43	9′ 10¼″	63	14′ 5¼″	83	19′ 0¼″
4	0′ 11″	24	5′ 6″	44	10′ 1″	64	14′ 8″	84	19′ 3″
5	1′ 1¾″	25	5′ 8¾″	45	10′ 3¾″	65	14′ 10¾″	85	19′ 5¾″
6	1′ 4½″	26	5′ 11½″	46	10′ 6½″	66	15′ 1½″	86	19′ 8½″
7	1′ 7¼″	27	6′ 2¼″	47	10′ 9¼″	67	15′ 4¼″	87	19′ 11¼″
8	1′ 10″	28	6′ 5″	48	11′ 0″	68	15′ 7″	88	20′ 2″
9	2′ 0¾″	29	6′ 7¾″	49	11′ 2¾″	69	15′ 9¾″	89	20′ 4¾″
10	2′ 3½″	30	6′ 10½″	50	11′ 5½″	70	16′ 0½″	90	20′ 7½″
11	2′ 6¼″	31	7′ 1¼″	51	11′ 8¼″	71	16′ 3¼″	91	20′ 10¼″
12	2′ 9″	32	7′ 4″	52	11′ 11″	72	16′ 6″	92	21′ 1″
13	2′ 11¾″	33	7′ 6¾″	53	12′ 1¾″	73	16′ 8¾″	93	21′ 3¾″
14	3′ 2½″	34	7′ 9½″	54	12′ 4½″	74	16′ 11½″	94	21′ 6½″
15	3′ 5¼″	35	8′ 0¼″	55	12′ 7¼″	75	17′ 2¼″	95	21′ 9¼″
16	3′ 8″	36	8′ 3″	56	12′ 10″	76	17′ 5″	96	22′ 0″
17	3′ 10¾″	37	8′ 5¾″	57	13′ 0¾″	77	17′ 7¾″	97	22′ 2¾″
18	4′ 1½″	38	8′ 8½″	58	13′ 3½″	78	17′ 10½″	98	22′ 5½″
19	4′ 4¼″	39	8′ 11¼″	59	13′ 6¼″	79	18′ 1¼″	99	22′ 8¼″
20	4′ 7″	40	9′ 2″	60	13′ 9″	80	18′ 4″	100	22′ 11″

TABLE 9–3 TOTAL HEIGHT OF COURSES USING 2¼-IN. BRICK AND ⅝-IN. MORTAR JOINTS

Courses	Height	Courses	Height	Courses	Height	Courses	Height	Courses	Height
1	0′ 2⅞″	21	5′ 0⅜″	41	9′ 9⅞″	61	14′ 7⅜″	81	19′ 4⅞″
2	0′ 5¾″	22	5′ 3¼″	42	10′ 0¾″	62	14′ 10¼″	82	19′ 7¾″
3	0′ 8⅝″	23	5′ 6⅛″	43	10′ 3⅝″	63	15′ 1⅛″	83	19′ 10⅝″
4	0′ 11½″	24	5′ 9″	44	10′ 6½″	64	15′ 4″	84	20′ 1½″
5	1′ 2⅜″	25	5′ 11⅞″	45	10′ 9⅜″	65	15′ 6⅞″	85	20′ 4⅜″
6	1′ 5¼″	26	6′ 2¾″	46	11′ 0¼″	66	15′ 9¾″	86	20′ 7¼″
7	1′ 8⅛″	27	6′ 5⅝″	47	11′ 3⅛″	67	16′ 0⅝″	87	20′ 10⅛″
8	1′ 11″	28	6′ 8½″	48	11′ 6″	68	16′ 3½″	88	21′ 1″
9	2′ 1⅞″	29	6′ 11⅜″	49	11′ 8⅞″	69	16′ 6⅜″	89	21′ 3⅞″
10	2′ 4¾″	30	7′ 2¼″	50	11′ 11¾″	70	16′ 9¼″	90	21′ 6¾″
11	2′ 7⅝″	31	7′ 5⅛″	51	12′ 2⅝″	71	17′ 0⅛″	91	21′ 9⅝″
12	2′ 10½″	32	7′ 8″	52	12′ 5½″	72	17′ 3″	92	22′ 0½″
13	3′ 1⅜″	33	7′ 10⅞″	53	12′ 8⅜″	73	17′ 5⅞″	93	22′ 3⅜″
14	3′ 4¼″	34	8′ 1¾″	54	12′ 11¼″	74	17′ 8¾″	94	22′ 6¼″
15	3′ 7⅛″	35	8′ 4⅝″	55	13′ 2⅛″	75	17′ 11⅝″	95	22′ 9⅛″
16	3′ 10″	36	8′ 7½″	56	13′ 5″	76	18′ 2½″	96	23′ 0″
17	4′ 0⅞″	37	8′ 10⅜″	57	13′ 7⅞″	77	18′ 5⅜″	97	23′ 2⅞″
18	4′ 3¾″	38	9′ 1¼″	58	13′ 10¾″	78	18′ 8¼″	98	23′ 5¾″
19	4′ 6⅝″	39	9′ 4⅛″	59	14′ 1⅝″	79	18′ 11⅛″	99	23′ 8⅝″
20	4′ 9½″	40	9′ 7″	60	14′ 4½″	80	19′ 2″	100	23′ 11½″

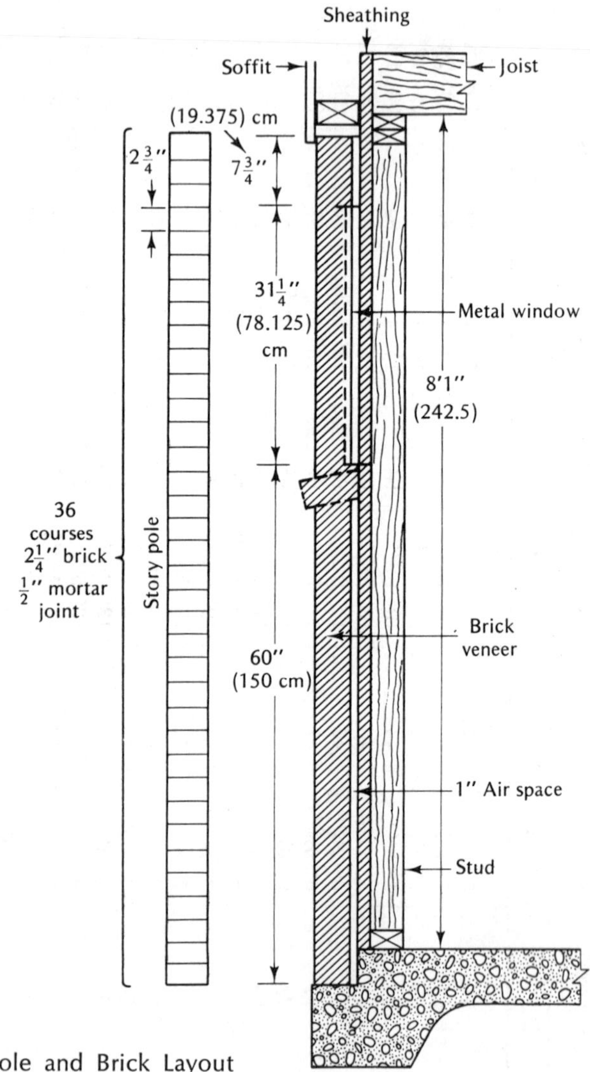

Figure 9–12 Story Pole and Brick Layout

Estimating Brick Quantities

There are several methods that can be used to calculate the number of bricks needed for a given job. The simplest uses either a quantity of 7 bricks per square foot of wall (single wythe) or 14 bricks per square foot of double wythe (8-in.-thick) wall. The procedure is as follows:

1. With the aid of the floor and elevation plans, determine the wall height and perimeter length.
2. Multiply the height value times the perimeter value to obtain the gross square feet (or meters).
3. Use either the window and door schedule or plans to calculate their total square feet (or meters), and subtract this value from the gross square feet (or meters).
4. Calculate the number of square feet (or square meters) space that receives no brick (e.g., paneling areas), and subtract this sum from the remainder of step 3.
5. Finally, to obtain the total number of bricks needed, multiply the net gross square feet by 7 for a single wythe or 14 for a double wythe (or square meters by 76 for a single wythe or 152 for double wythe).

Example 1: Calculate the brick veneer needed on a frame house whose height is 8 ft 3 in. and perimeter is 200 linear feet. The schedule indicates 534 square feet of window, door, and garage door opening.

1. Wall height	8 ft 3 in.
2. Wall perimeter length	200 ft 0 in.
Gross square feet	1650 ft 0 in.
3. Less window and door openings	− 534 ft 0 in.
Net gross square feet	1114 ft 0 in.
4. Not used	0 in.
Total net gross square feet	1114 ft 0 in.
5. Use 7 bricks per square foot	× 7
Total number bricks needed	7798

By using 7 (or 14, as the case may be), an error is included that results in sufficient brick to cover breakage.

More refined estimates can be obtained by using a more involved formula. It involves determining the number of courses to be laid, and, of course, the thickness of the joint plays a significant part in this calculation. Next, the brick's length plus one vertical joint thickness is used to determine the total number of bricks needed for one course around the perimeter. Then the two values are multiplied to find the gross number of bricks. Next, the number of bricks equal to the sum of all openings—windows, doors, garage doors, and paneling—is subtracted from the gross sum. Finally, 5 to 10 percent is added to the quantity of bricks figured, to allow for breakage.

Bricklaying Patterns

Eight variations or patterns of bricks are illustrated in Figure 9–13. Three are frequently used when brick-veneering a house: the running, the one-third running, and the stack. Not one of these three patterns uses headers; in fact, their installation does not permit the use of headers. Therefore, wall ties are used to secure the bricks to the wall.

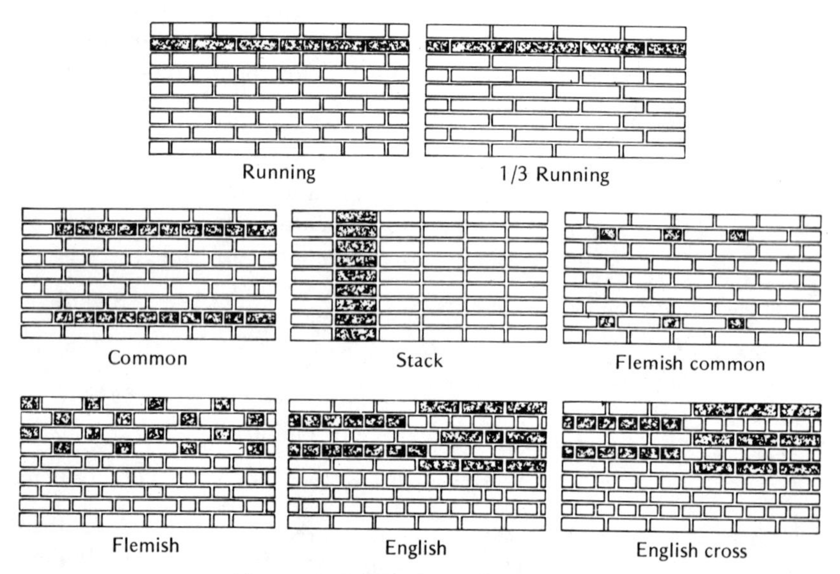

Figure 9–13 Brick Wall Patterns

The three patterns just listed and the remaining five may be used for a full brick wall. To aid in distinguishing each pattern, a concise definition is provided, as follows:

a. *Running.* The running pattern consists of stretchers only. Notice that vertical joints are centered above and below the middle of the alternate course.
b. *One-third running.* The one-third running pattern uses only stretchers, but the vertical joints fall one-third of a brick's length away from its end.

c. *Common.* The common pattern consists of a first course of continuous full headers, followed by five courses of stretchers and another course of continuous full headers.
d. *Stack.* The stack pattern has the least strength and as a rule cannot be used as a load-bearing wall. Notice that all vertical joints are aligned.
e. *Flemish common.* The flemish common pattern consists of a course made up of alternate stretchers and headers, followed by five courses of stretchers and another course of alternate stretchers and headers.
f. *Flemish.* The flemish differs from the flemish common in that each course has a repeat pattern of a header and a stretcher. Further, each header is centered above and below the middle of the stretcher brick.
g. *English.* The english pattern consists of alternate courses of headers and stretchers where the headers are centered over the stretcher joints. Notice that every fourth vertical joint alignment is made on the stretcher joints only.
h. *English cross.* The english cross is similar to the english except that all stretcher joints line up vertically.

Joint-Striking Techniques

Both horizontal and vertical joints are struck on all brick walls as a rule. Striking provides a uniform appearance and creates a surface that sheds water easily. It should be done within a period of several hours after the bricks have been laid. If too much time elapses, considerable effort is needed to strike the joint, and sometimes mortar is discolored because of the tool being used.

A variety of shapes can be made in the mortar, and Figure 9–14 illustrates some of them. Notice, too, the tool used to make the joint.

The *flush joint* is easily made with a brick trowel; however, its watertightness is limited, in that the compression of the mortar made by a striking tool is absent.

The *raked joint* falls in this category, too. In addition, a ledge is formed where water and mildew can remain. The joint is attractive, though, and is useful in creating shadows.

The *concave joint* is the most common. It is weathertight because the tool used compresses the mortar. The other joints shown are variations of the concave and retain the watertightness characteristic.

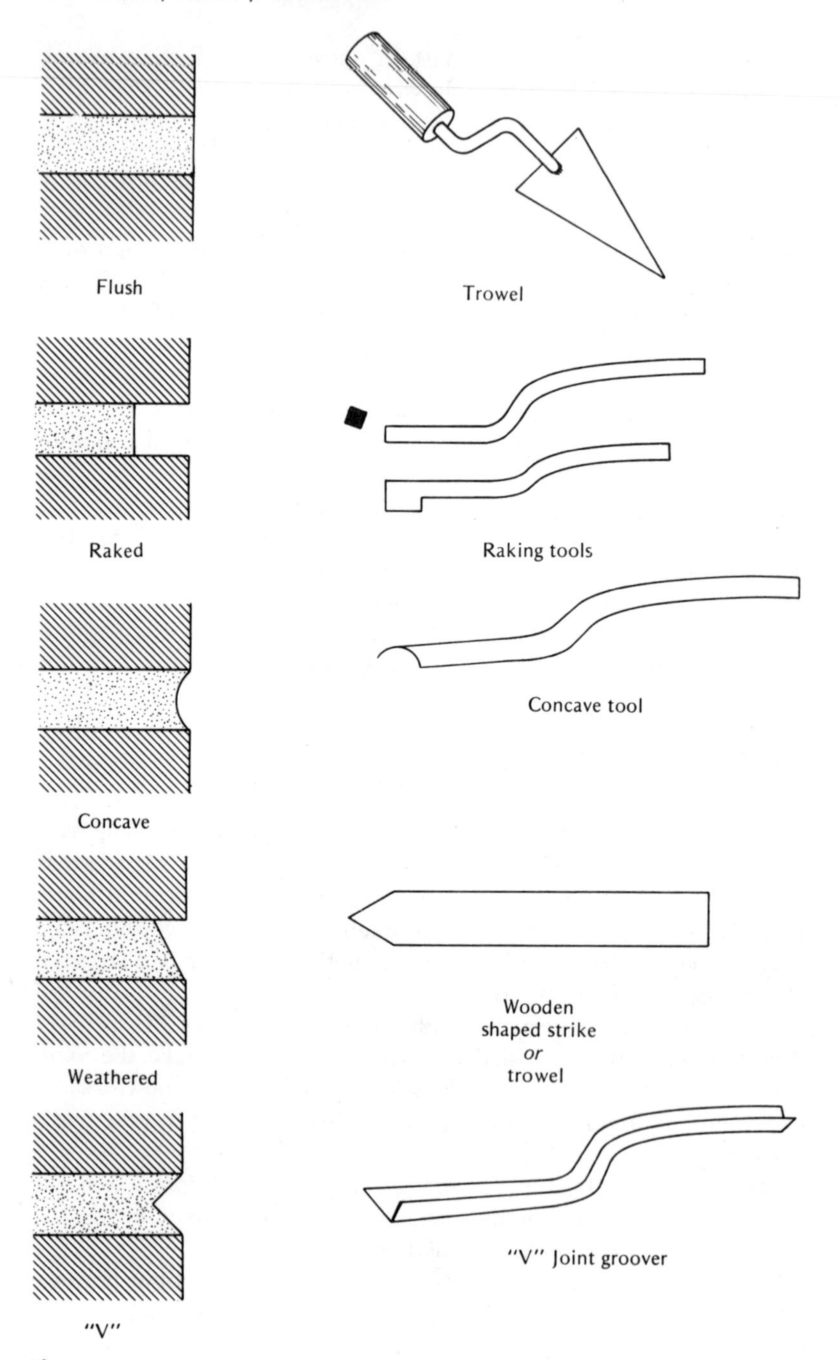

Figure 9–14 Striking Joints, Their Names, and the Tool Used for the Job

INSPECTION OF THE BRICK JOB

Inspection	*Satisfactory/N.A.*	*Unsatisfactory*
Was a proper foundation laid?	____	____
Was a course layout made?	____	____
Was the proper type of mortar used?	____	____
Were proper ties used and in sufficient quantity?	____	____
Are all joints in proper alignment?	____	____
Were rowlock sills tilted to allow for proper drainage?	____	____
Were large-enough lintels installed?	____	____
Were all exposed joints struck uniformly?	____	____

QUESTIONS

1. Define the following terms.
   a. Bonded.
   b. Header.
   c. Lintel.
   d. Rowlock.
   e. Stretcher.
   f. Wythe.
2. What does the term "veneering" mean in bricklaying?
3. How does brick veneering remain attached to a wall?
4. Which drawing shows brick-veneering details?
5. Why include weep holes in a veneered wall?
6. What type of brick arrangement is used at windowsill level? At the top of the wall?
7. Explain the purpose of a lintel.
8. Which bricks above the lintel are supported by the lintel?
9. How far above the lower lip of the frieze board must the top course of brick extend?

10. In a solid brick wall, how many wythes are usually laid?
11. How does a header course add strength to a double-wythe wall?
12. Can metal ties be used to unite wythes?
13. Explain how to install lintels on a double-wythe brick wall over a window opening.
14. What is a standard mortar mix made from?
15. What is the better brick joint, 3/8 in. or 5/8 in.?
16. Is a story pole used during bricklaying?
17. What is the roughly estimated number of bricks per square foot?
18. Explain the differences among the running, the one-third running, and stack patterns of bricklaying.
19. Name and describe three types of joint-striking techniques.

Chapter 10

The Brick Barbecue

Buttered the small end of a brick that has a quantity of mortar placed onto it (e.g., the act of buttering a brick's end with mortar).
Closure brick a partial brick that is cut to fit into a place to complete a course.
Double wythe two vertical columns of brick, in this chapter united by grouting and interlocking headers.
Furrow a spreading of mortar after placing it onto a course of brick; similar to furrowing a field for planting.
Grout mortar or thinned mortar used as a binder between wythes.
Header course a course of bricks where each brick is laid at right angles to the stretchers.
Mason's line a strong, usually white, ⅛-in. cord or line used to establish the alignment of bricks.

OBJECTIVES—INTRODUCTION

The brick barbecue is normally built in the backyard or just off the patio. Many questions arise as to its shape, construction requirements, and material needs. Books and magazine articles on the subject provide many clues and ideas. Sometimes you will find the ideal design, sometimes not. What really complicates the situation is not the construction but the purpose. If the barbecue is to be used only for outdoor cooking, the design will be simple. Only when you have plans for making a smoker barbecue, for example, does the design become complex.

When working with the simply designed ones, such as the example detailed in this chapter, the major concerns are: size; type of heating: gas, electric, or wood; types of materials to use; and design features. This then leads to the objectives for this chapter, which are: *to be able to formulate a plan for construction of a backyard barbecue*, and *to be able to translate the barbecue plan into actuality.*

The first step is to read the plan for the barbecue so that its details are familiar. Then they may be applied to a plan drawn to satisfy individual needs or local situations. From this understanding the construction of the barbecue can be defined in terms of task sequencing and applications. If the sequence is adhered to, the probability of success is greatly improved. The data developed in Chapter 9 should be correlated with this chapter's data.

READING AND UNDERSTANDING BARBECUE PLANS AND SPECIFICATIONS

Skilled draftsmen may be able to render a set of blueprints for a house or even a simple barbecue by just understanding the overall requirements. Actually, they mentally construct a total image of the whole project and include its details. The beginner or apprentice mason or the homeowner probably cannot do this. Therefore, a pictorial or rough sketch should be made or obtained prior to developing and understanding plans.

Figure 10–1 is a pictorial sketch of the barbecue used as the example in this chapter. Notice that it is shown in a setting within a

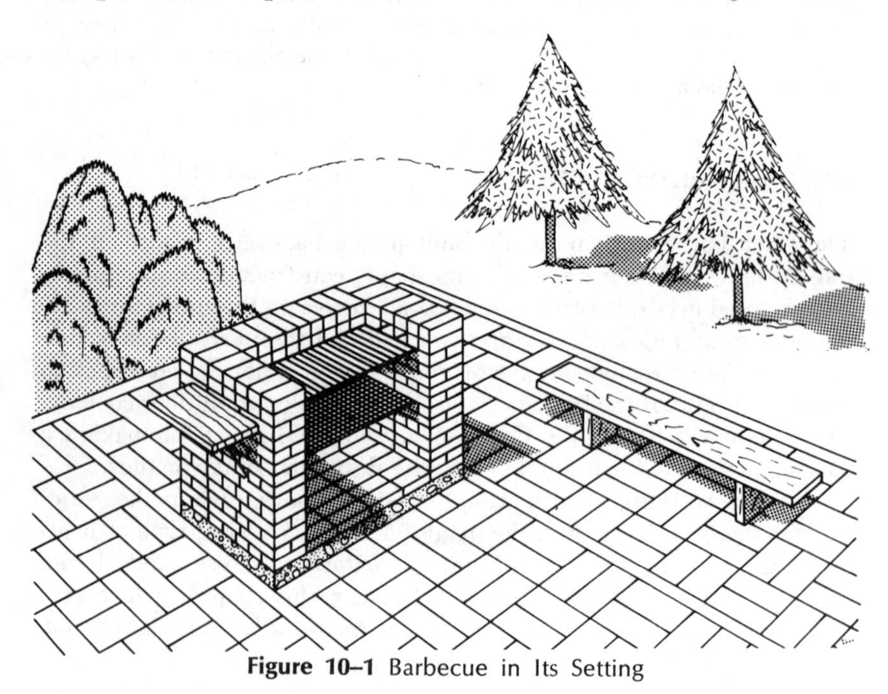

Figure 10–1 Barbecue in Its Setting

brick patio, bordered on the back by shrubs of varying height; it has a bench as an accent feature. The barbecue itself is drawn to include many of the necessary and desired features:

a. A serving and preparation shelf anchored on one side.
b. A fixed height for a charcoal grate.
c. A variable-position cooking grill.
d. A concrete slab.
e. Brick in double wythe forming the walls.
f. A beginning and top header course.

With this much detail pictured, the plans are easier to understand. Figure 10–2 is the set of plans that is needed for construction of the barbecue. The set of plans consists of three parts: the *foundation* and top view, the front-view *elevation,* and detail elevation A–A.

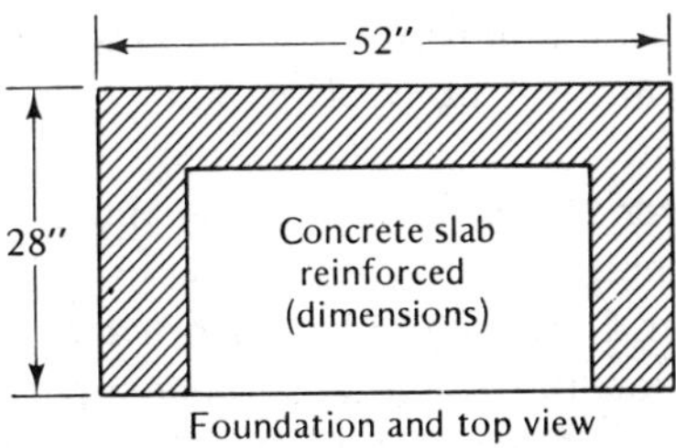

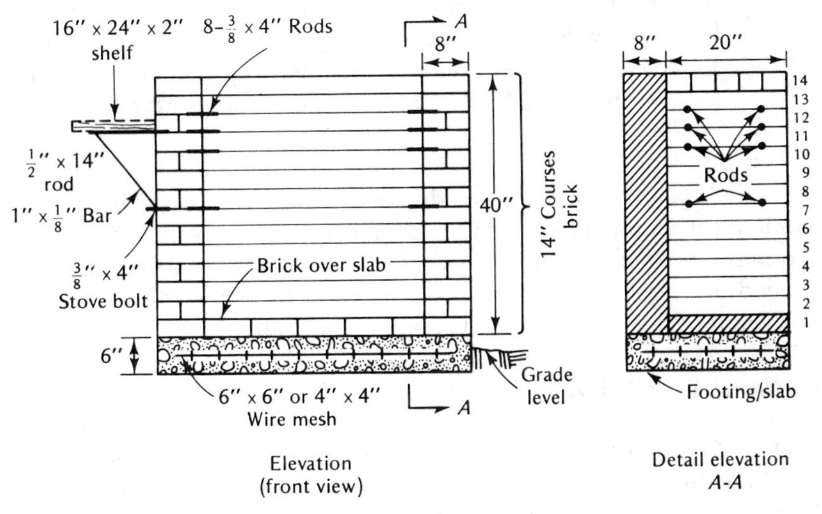

Figure 10–2 Barbecue Plans

The Foundation

The foundation plan shows that the overall dimensions of the barbecue are 28 in. front to back, or deep, by 52 in. long. This rectangle must be formed with either 1 × 6 or 2 × 6 stock, and all four corners must be right angles.

Notice that the drawing indicates that the bricks (shaded area) are to be laid flush with the outer edge of the concrete. This feature, plus the use of standard 8-in.-long bricks develops a grill area 20 in. deep by 36 in. long, or a total of 720 in.^{2}

The Elevation

The elevation plan includes the 6-in. thickness of the foundation. It also shows the following data:

1. Some portion of the foundation is to be above grade level.
2. Wire mesh, 6 in. × 6 in. or 4 in. × 4 in., should be used for foundation requirement.
3. Bricks may be laid over the slab within the pit area to add character to the barbecue.
4. Rods of ⅜-in. steel cut 4 in. long are embedded between the courses of brick to support the fire grate and cooking grill.
5. Two ⅜ in. × 4 in. stove bolts are to be installed with threads outward to bolt the shelf brace in place.
6. Two ½ in. × 14 in. rods are to be installed between courses of brick that will support the working shelf.
7. By count there are 14 courses of brick and the way the joints are shown, the method of bricklaying is one using solid double wythe.

A special note must be made about data item 1, which refers to grade level. In especially cold climates it may be necessary to include footings under the slab. They would reach down farther into the ground and provide a better, sound anchor. If such is the case, refer to Chapter 2 for footings and Chapter 7 for slabs.

Sectional Detail A-A

Section A–A (Figure 10–2), which is a detail elevation plan taken from the elevation plan, provides additional data needed for construction. First, it shows the approximate positions of rods that will support the grills. It also shows the number of courses of brick as 14.

Specifications and Materials List

Specifications for Barbecue The specifications should be as simple as possible yet contain data not readily apparent or not suited for inclusion in plans. For example, the specifications for the barbecue could include:

1. Use 2500-psf-compressive-strength concrete with coarse aggregate limited to ¾ in.
2. For mortar use 1 part cement, 4½ parts fine sand, and ½ part fire clay.
3. Estimate 50 bricks to a sack of dry-mix mortar.
4. Set the first course of brick using headers.
5. Use a double-wythe system for walls of the barbecue with grouting or mortar between wythes.
6. Use a header course for the final or top course.
7. Install rods and bolts so that more than 50 percent of each rod is embedded in mortar. (Exceptions are the 14-in. rods.)
8. Strike all joints before the mortar sets but after it begins to set.
9. Dampen all bricks at least 8 hr before using.

Estimate of Materials for Barbecue The estimate is developed by calculating the various needs. First the foundation needs are determined. Next the number of bricks are determined; then mortar requirements can be calculated. Finally, steel rods, grills, and the work shelf should be listed. The following example reflects the needs of the barbecue shown in Figure 10–1.

Foundation

5 ft^3	concrete (28 in. × 52 in. × 6 in.)
8 ft^2	wire mesh
14 linear ft	1 in. × 6 in. stock (form)
8 ft	1 in. × 2 in. stock (stakes)

Barbecue

400	hardened bricks	2	⅜ in. × 4 in. stove bolts
½ yard	sand	1	2 in. × 6 in. × 4 in. shelf
3 bags	cement	4	1¼ in. No. 6 RH screws to screw shelf onto rods
1 bag	fire clay	2	20 in. × 1 in. × ⅛ in. steel bars (for brace)
32 in.	⅜-in. steel rods	1	18 in. × 36 in. cooking grill
28 in.	½-in. steel rods	1	18 in. × 36 in. fire grill

Tools Needed

- Trowel
- Mason's hammer
- Brick set
- Mortarboard
- Chalk line
- Story pole (optional; Figure 10–3)
- Claw hammer
- Spirit level
- Shovel
- Hoe
- Nails

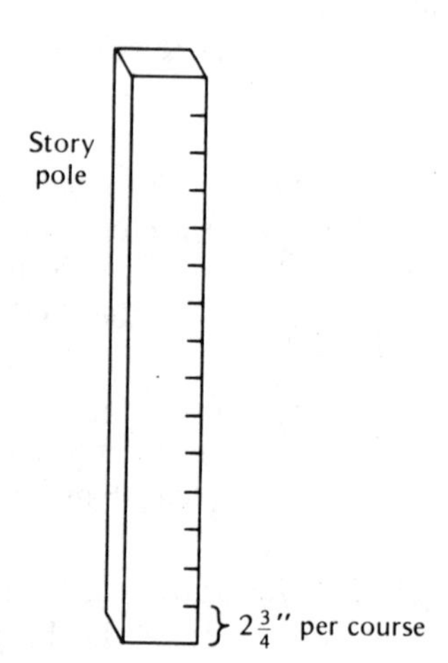

Figure 10–3 Story Pole

CONSTRUCTING THE BARBECUE

The concrete foundation must be poured and allowed to set and harden before laying up the bricks. Next, the actual length of the bricks should be measured with a ruler to determine that none are longer than 8 in. If any longer than 8 in. are used, they could alter the interior dimensions of the grill, making it smaller than 36 in. The cooking grill and fire grate would not fit. So it is very important that the interior dimensions be a minimum of 36 in.; however, 36½ in. is acceptable.

Laying the Base and Intermediate Courses of Brick

The first course of brick should be laid dry and in header formation, as Figure 10–4 shows. Notice that one closure brick (*) is needed to complete the header course along the back wall. When placing the dry bricks, allow a ⅜-in. to ½-in. space for mortar. Next mix a batch

Figure 10–4 Header Course

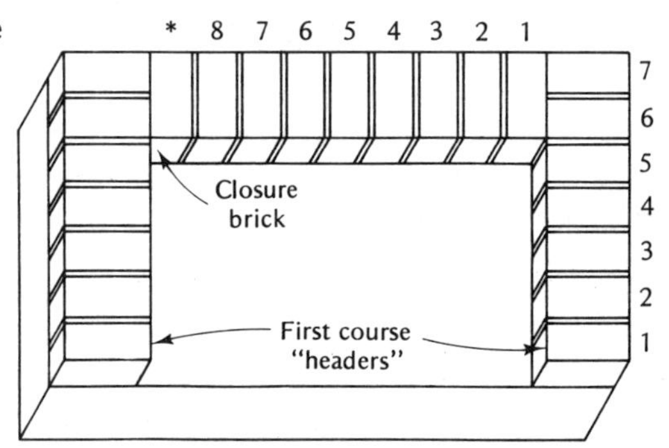

of mortar, and mortar each header into place. (Verify the 36-in. interior dimension.)

The normal routine to follow is to lay up the corners, then lay up the intermediate bricks. To help in understanding which way to lay up each course, Figure 10–5 shows the odd- and even-numbered courses numbered 2 through 13. Course 14 is a header course duplicating course 1.

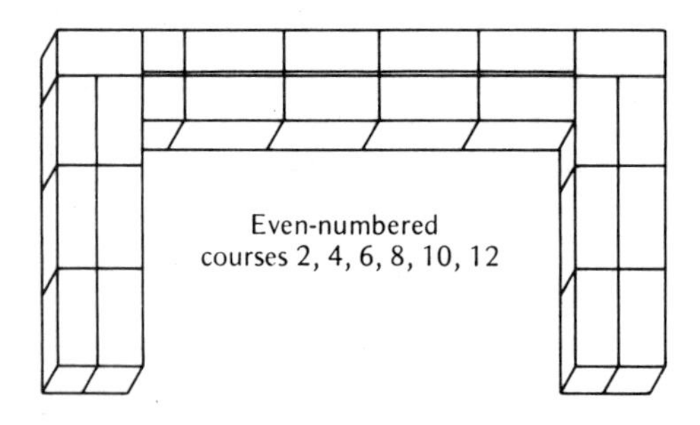

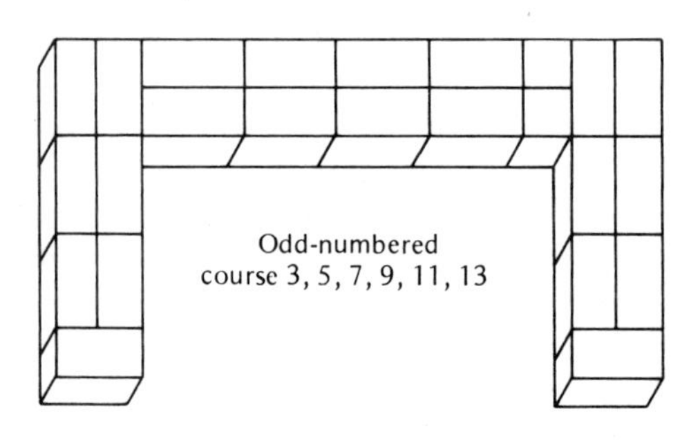

Figure 10–5 Course Layout

Installing Bolts and Rods

After laying up the seventh course of bricks cut and install the ⅜ in. × 4 in. rods that will be used to support the fire grate and the two ⅜-in. stove bolts that will be used to secure the brace.

After laying up the tenth course, install the ⅜-in. rods for the cooking grill and repeat the procedure after laying up the eleventh and twelfth courses. Also, on the twelfth course install the two 14-in. rods that will support the work shelf. Laying the thirteenth course will secure all the rods in position.

Capping the Top and Bricking over the Concrete Foundation

The fourteenth or top course of brick is to be laid header style. This course must be made from solid, unperforated bricks because the flat surface is the finish surface. The proper positioning of the headers is shown in Figure 10–3 and duplicates the first course of bricks. Since three surfaces of these bricks ultimately show, sufficient mortar must be used so that striking can be done on all three surfaces.

After the top course is laid, bricks should be mortared on top of the concrete within the barbecue. The pattern selected can be randomly made, where the least amount of cutting is done. Before applying mortar to the concrete surface, lay and cut all bricks needed. Then mortar all in place.

Striking and Cleaning

Any of the striking methods shown in Chapter 9 can be used on the joints. They should all be made uniform, and vertical joints are usually completed before horizontals. The top surface joints may be left flush so that a flat, even surface is available for setting plates, salt and pepper shakers, soft drinks, and the like.

Cleaning is done with a solution of 1 part muriatic acid to 9 parts water. First spray the brick surface to be cleaned with water and allow it to penetrate. Then brush on the mixed solution with a 3- to 1-in. paint brush, rubbing well onto the brick surface. Rinse off the solution with clear water so that no acid remains on any part of the surface. (*Note:* Acid that is not washed off causes discoloration.)

Scoop and trowel full
of mortar
Lay it on the course
Spread it along
Furrow the mortar
Trowel off excess
Butter brick
Line
Buttered brick being
laid in place
Pushing brick in line and
cleaning excess mortar
Strike along brick
surface for flush joint
Continuously
verify
vertical
plumb
with
spirit
level

Figure 10–6 Laying a Brick

Task Description and Summary

Each brick is laid in a bed of mortar. So it is necessary to first place a trowel or several trowelfuls of mortar where the brick is to be placed. The mortar should be at least ½-in. thick and furrowed to make it spread toward the outer edges of the course below. The end of the brick being placed is buttered and then placed onto the mortar. It is tamped into final position with a trowel or hammer. Excess mortar is struck off with the trowel in an *upward* sweeping motion to prevent spillage on bricks below. This places the excess mortar onto the trowel and it should be added as a butter to the exposed end of the brick just laid. Figure 10–6 shows the step-by-step sequence just described. The level should be placed across the top of the brick first to determine accuracy, or its top edge must just touch a straight line. Its vertical alignment must be verified and maintained by using the level and adjusting the brick's position.

As each course is being laid up, mortar or grout must be filled between the wythes. The mortar being used for brick laying may be somewhat stiff, so a softer mixture can be made for grouting, and it can be run into the joint.

INSPECTION OF THE BARBECUE

Inspection	*Satisfactory/N.A.*	*Unsatisfactory*
Was a proper foundation laid?	______	______
Was the first course of brick laid with square corners?	______	______
Was sufficient space allowed for the grills to fit?	______	______
Was each course laid up plumb and level?	______	______
Were all bolts and rods properly located and soundly secured?	______	______
Were the cap bricks laid in header fashion and struck flush?	______	______

Inspection	*Satisfactory/N.A.*	*Unsatisfactory*
Were all joints struck?	______	______
Were bricks cleaned and cleaning agents washed away?	______	______

QUESTIONS

1. What does "buttering" a brick's end mean?
2. Explain where a closure brick is used.
3. What is grout and where is it used?
4. List the details that should be found in the plans of a barbecue that are needed by the mason.
5. Is the barbecue explained in this chapter constructed with one or two wythes of brick?
6. How will the fire grate be supported?
7. What advantage is there in making a trial setting of the first course of brick without mortar?
8. What chemical is used to clean brick surfaces after they have been stained with mortar?
9. Why does the mason furrow his mortar bed?

Chapter 11

Brick Patios and Sidewalks

Felt 15-lb (or heavier) tar-impregnated paper used for insulation.
Grade the final or top surface of a structure, such as the top surface of a patio. The reference point from which other heights are defined.
Pavers bricks in numerous sizes and shapes that are used in constructing sidewalks, patios, and driveways.
Tar paper the same as felt paper.
Toenailing the nailing of the end of one member to the side of another with nails used at angles or toed.

OBJECTIVES—INTRODUCTION

The construction of a brick patio or brick sidewalk involves two concepts: architectural style and construction requirements. The idea of using brick to construct a patio or sidewalk presents the builder with a variety of opportunities to create artistic forms. Many patterns can be developed, and two, the circular and basketweave pattern, are shown in Figure 11–1. Later in this chapter other patterns are shown and explained; however, this leads to the first objective for the chapter: *to be able to create a design with brick in a patio or sidewalk.*

Before or concurrently with the selection of the arrangement of bricks, a proper method of installation must be selected from several alternatives. The method of construction may depend upon such factors as sloping, drainage, anchoring, subsurface formation, and techniques needed to install the bricks or pavers. This leads to the second objective of this chapter: *to include sound construction methods when planning and constructing a brick patio or sidewalk.*

Figure 11–1 Circular and Basketweave Brick Patterns

READING AND UNDERSTANDING LANDSCAPE AND ELEVATION PLANS FOR PATIO AND SIDEWALK, INCLUDING SPECIFICATIONS

The three most significant aspects one must consider when reading and understanding plans for a patio or sidewalk are proper location, proper grade and slope, and overall appearance or design. Three lesser requirements include defining the condition of the soil under the patio or sidewalk, the base needs for brick stability, and the type of brick to use.

Proper Location

A brick patio or sidewalk could easily be substituted for a concrete patio or sidewalk. Recall the landscape plans in Chapter 7 (pp. 96–99), shown in Figures 7–1 and 7–2. Figure 7–2 is reprinted here as Figure 11–2 so that it can be used in determining patio and sidewalk needs. This time, however, the plan will be used for the construction of both patio and sidewalks using brick as the face materials. To be sure that

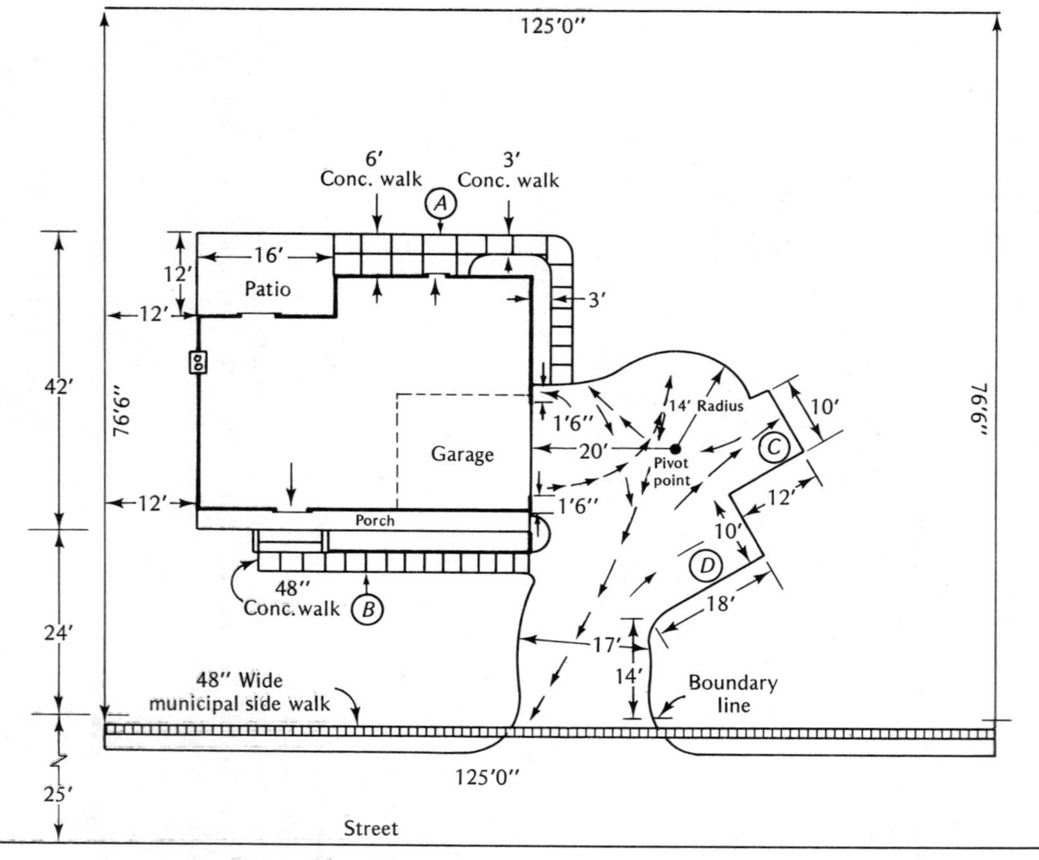

Figure 11–2 Sidewalks and Driveway Measurements

a clear understanding exists of the positions of the patio and sidewalks in Figure 11–2, their locations and sizes are listed:

1. The patio is attached to the house at the rear of the house and is 12 ft × 16 ft (3.6 m × 4.8 m).
2. The rear sidewalk is 6 ft (1.8 m) wide from the patio to the point where it narrows to 3 ft (0.90 cm).
3. The front sidewalk is 4 ft wide (1.2 m) from the front steps to the driveway.

The location data and parameters outlined on this plan detail rectangular quantities and conform quite easily to standard brick measurements, since brick pavers are readily available in 4 in. × 8 in. dimension. The next aspect to consider is the proper grade and slope for the patio and sidewalks.

Proper Grade and Slope

Elevation data need to be defined regarding the amount of slope on the patio. Industry standards dictate that patios shall have a slope of ¼ in./foot of run to allow for proper drainage away from the house. Considering that the patio shown in Figure 11–2 is 12 ft wide, the total slope is 3 in.

$$\frac{1/4 \text{ in.}}{12 \text{ in.}} \times 144 \text{ in.} = \frac{36 \text{ in.}}{12 \text{ in.}} = 3 \text{ in.}$$

Translating this requirement into an elevation plan results in a drawing, as Figure 11–3 shows. This drawing clearly details the total slope of the 12-ft-wide patio. Notice that the patio's top surface starts at the house, one brick's thickness below the finished floor level. This arrangement results in a small downstep from house to patio, allows weep

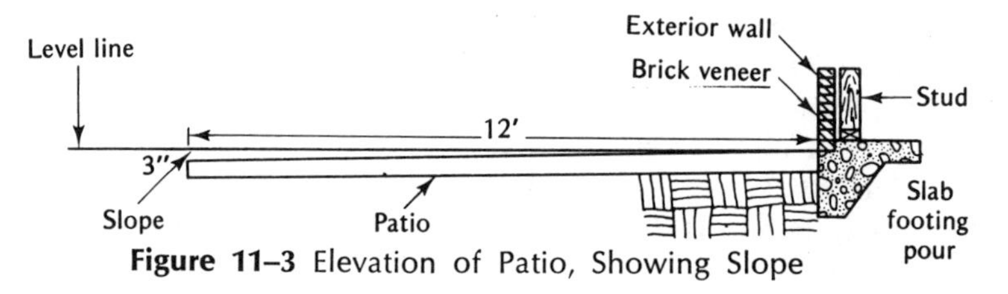

Figure 11–3 Elevation of Patio, Showing Slope

holes in the veneer to function, and prevents heavy rains from flooding the house.

All patios and sidewalks should slope, for several reasons. The first has already been defined as a need to carry water away from the house. Another reason for sloping the surface is to minimize the penetration of water into the brick and mortar, if mortar is used. By reducing the time that water remains on a brick surface, absorption is reduced and shrinking and swelling are minimized. Standing water on the bricks' surface also creates problems during freezing and thawing, so this should be minimized by using the sloping technique. An equally effective technique in eliminating the chance of standing water is to plan and construct a crown in the sidewalk (Figure 11–4). With location and grade and slope defined, the overall appearance can be developed.

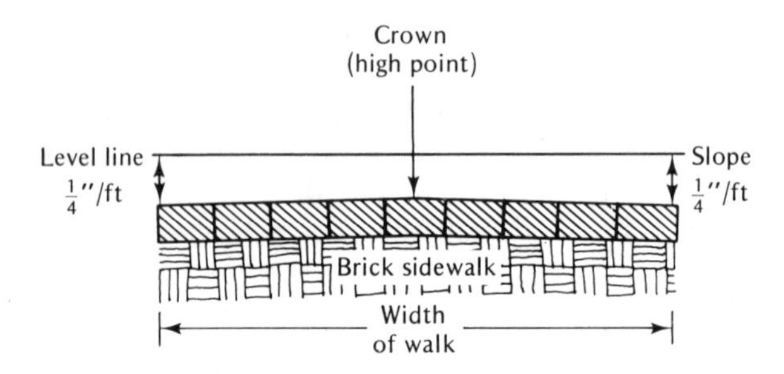

Figure 11–4 Crown in a Brick Walk to Shed Water

Defining Overall Appearance

The overall appearance of a brick patio or sidewalk includes several important considerations: the architectural pattern (style) desired, the border pattern, and any grids made from materials other than brick. The patterns for rectangular proportions are shown in the eight examples in Figure 11–5.

Patterns A through D and G can be laid with 4 in. × 8 in. bricks and very little joint spacing exists. Patterns E, F, and H can be laid tightly with all sizes of brick since each pattern is a running bond variation. Bricks with sizes other than 4 in. × 8 in. may be used for patterns E, F, and H; however, a mortar joint may be developed if they are used with the other patterns. Another pattern that can be used is shown as circular in Figure 11–6. Because of the rectangular shapes of

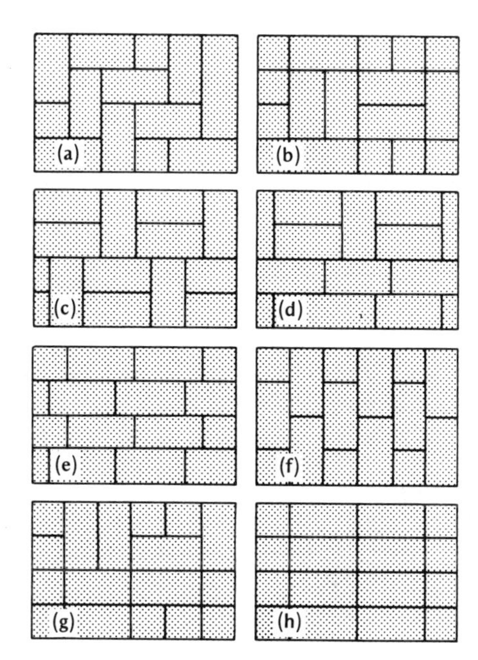

Figure 11-5 Eight Brick-Paving Patterns for Patio and Sidewalk *(Courtesy Brick Institute of America)*

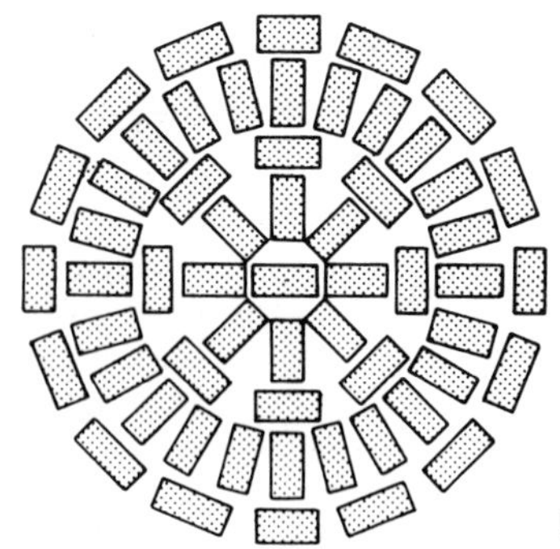

Figure 11–6 Circular Pattern of Brick Patio

each brick, this pattern causes fairly large mortar joints near the center and smaller joints as the radius becomes larger.

Keeping pavers in place requires a permanent type of border or mortaring to a base. One method is to use redwood 2 × 4s. These members should be installed such that they do not shift or raise. Stakes may be driven alongside and nailed to the 2 × 4s (Figure 11–7). Or iron rods 16 in. to 18 in. long by ⅜ in. in diameter may be driven through previously drilled pilot holes (Figure 11–8). Or the 2 × 4s may

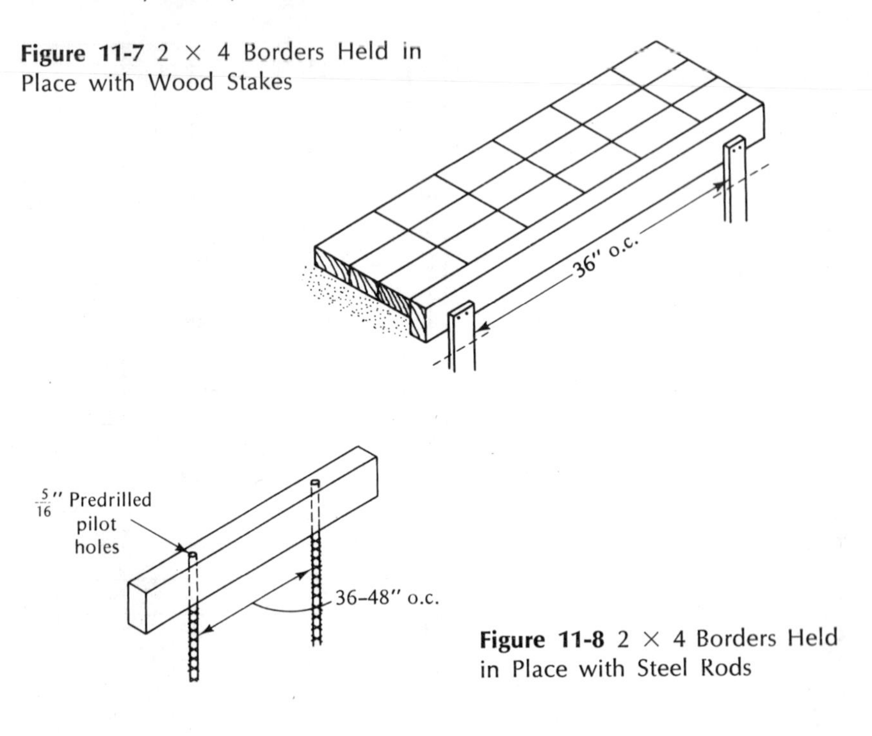

Figure 11-7 2 × 4 Borders Held in Place with Wood Stakes

Figure 11-8 2 × 4 Borders Held in Place with Steel Rods

be embedded into a concrete footing similar to edge bricks (soldier course) shown in Figure 11–9.

Several other possibilities that exist for confining pavers are bricks installed in soldier fashion with their lower half embedded in a concrete footing, or brick curbing also embedded in concrete. Both design methods firmly secure the border bricks, thereby restricting any paver movement. Finally, a base of concrete or mortar may be used to hold pavers in place. In summary, if no border support is included when constructing a mortarless patio or walk, pavers continually shift. The edge pavers separate, tilt, and become uneven. This causes a serious safety hazard, in addition to a deteriorating appearance. But if a border is included, the pavers remain in place, and if a border is not desired, the bricks must be secured to a concrete or mortar base.

Defining Soil Conditions

Any sidewalk or patio may have to be built over sandy soil, clay soil, gravel mixed soil, backfilled soil, or marshy soil. Each type of soil creates a different set of problems. Sandy soil, for example, readily fills with water and expands. This causes uneven vertical disturbance to

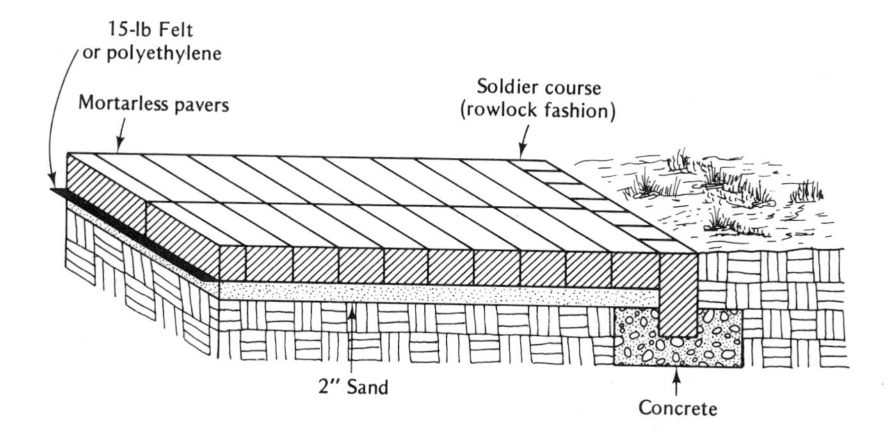

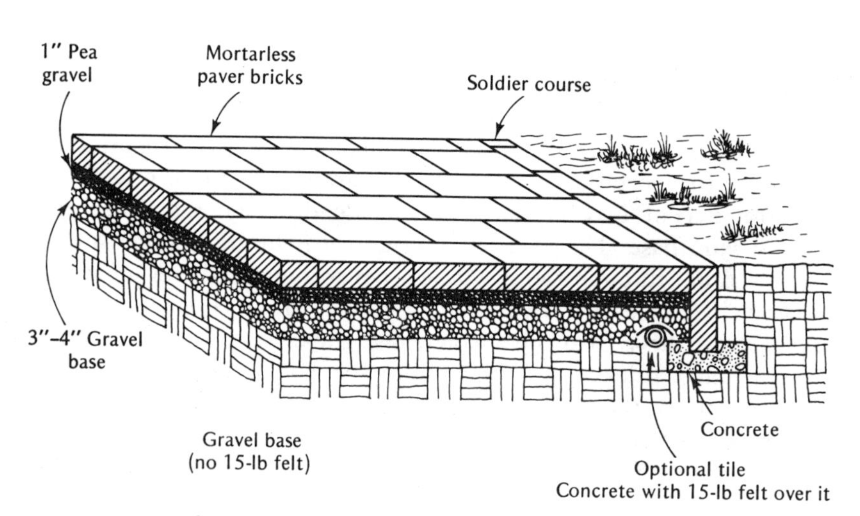

Figure 11–9 Types of Bases for Brick Pavers

sidewalk or patio. Clay soil, in contrast, does not absorb water readily and thereby creates a drainage problem, and when it is wetted, it causes a sliding condition. Gravel or rocky filled soil is fine if virgin, but if used as fill, it contains many air pockets that will fill over the years with small aggregate and sand, causing settling and an uneven sidewalk or patio surface. Backfilled or marshy soils also settle with time and must be reckoned with.

It therefore becomes important to study the ground where the patio or sidewalk is to be located. One should take several soil samples to determine a subgrade soil content. Then, if warranted, drainage can be included in the plan of construction. It may also be revealed that the sidewalk should serve a dual function (1) to walk upon, and (2) to carry water away from developed areas. Several methods for this are discussed in the next section.

A need to overcome a soil problem results in several suggestions. The use of gravel from large to pea installed several inches thick over sand minimizes the effect of sand's expansion. Clay should be dug out and removed to a depth of 6 in. and gravel backfilled and compacted. Disturbed gravel and rocky soils should be filled with pea gravel and sand and thoroughly wetted to wash these materials into all crevices. Backfilled soils must be compacted thoroughly; marsh soils should be removed and backfill installed and compacted. When it is not possible to remove the marshy soil, gravel should be used to stabilize the soil, or a concrete base under the pavers should be considered.

Below Paver (Base) Needs

Figure 11–9 illustrates the two most common methods of preparing the subsurface for mortarless pavers. One illustrates the use of gravel, and this arrangement may or may not include drainage tile. The other shows a sand base. The idea of using gravel is to allow some surface water to drain through the mortarless paver joints and run off within the gravel. Notice that two sizes of gravel are recommended. The larger is used first; then the smaller, pea gravel is used at a depth of 1 in. The paver bricks rest directly on top of the pea gravel.

In contrast, the sand-base method includes 2 in. of washed, clean sharp sand. The kind of sand used in masonry or concrete is best suited for this purpose. Since the sand is used to develop an even surface and not for drainage, it should be covered with a thickness of 15-lb felt (tar paper) or plastic polyethylene. Either material serves two purposes: (1) it restricts the surface rainwater from entering the sand base, and (2) it inhibits growth of grasses and weeds that would otherwise tend to grow between pavers.

Figure 11–10 illustrates the concrete base that is used when pavers are mortared in place. The base may be established in one of two ways. In the first, the concrete is poured into the formed area, and while still plastic, the pavers are laid with mortar between their joints. The second way is to prepare the base, allow it to dry, and then lay the pavers with mortar.

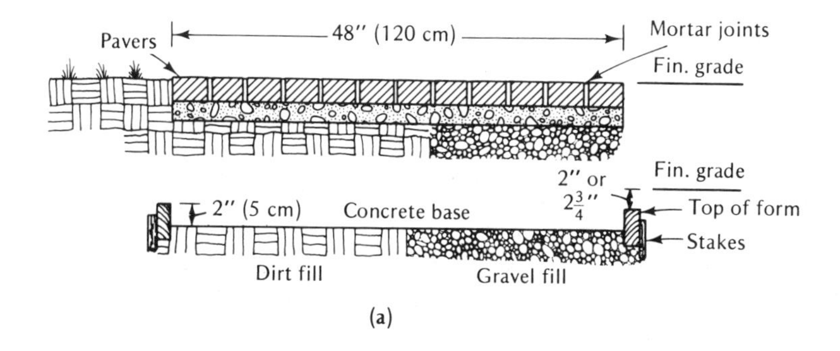

(a)

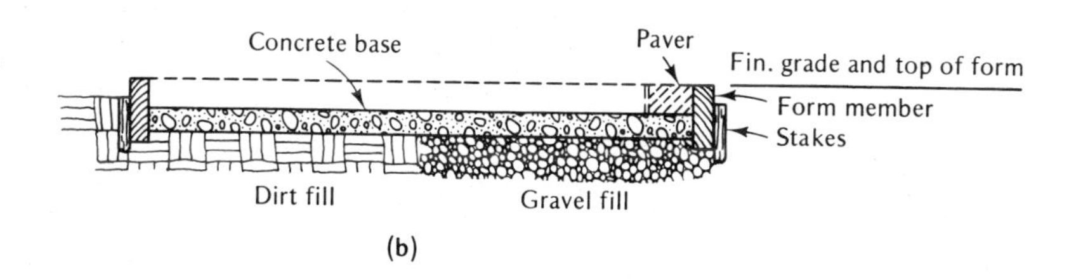

(b)

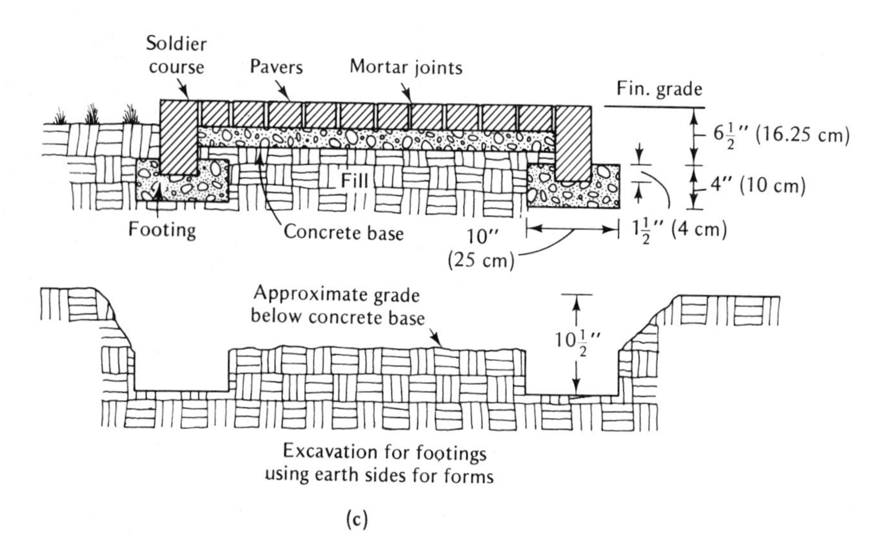

(c)

Figure 11–10 Base when Mortaring Pavers

Type of Bricks to Use

The discussion in this chapter has for the most part centered around the standard 4 in. × 8 in. paver brick. Whether this size or others are used, the type of brick should be *grade SW* (severe weather), in accordance with ASTM C216 or C62. Generally, the paver brick is solid rather than cored (one with three or more holes) because its widest surface is usually placed up, and the holes would detract from the appearance. The color of the brick is not a guide to its grade, since American brick manufacturers are able, by a combination of minerals and manufacturing processes, to develop all manner of appearances in many grades.

Table 11–1 lists 25 sizes and three shapes of brick pavers. It also provides the number of square inches per face area as well as the number of pavers needed per square foot (last column). This table should be a valuable asset in calculating the number of bricks for a patio or sidewalk. As an example, recall that the patio in Figure 11–2 is measured as 12 ft × 16 ft. If a 4 in. × 8 in. paver (the top line in Table 11–1) is used, the total quantity needed is 864 plus 5 percent for breakage and waste. For example:

$12 \text{ ft} \times 16 \text{ ft} = 192 \text{ ft}^2$
$192 \text{ ft}^2 \times 4.5 \text{ pavers/ft}^2 = 864$
$5\% \text{ waste} = 0.05 \times 864 = 43$

Total bricks needed 907

(*Note:* Since bricks are usually ordered in 500- or 1000-lot units, the order would be for 1000 bricks.)

In metric:

$12 \text{ ft} = 3.6 \text{ m}, 16 \text{ ft} = 4.8 \text{ m} = 17.28 \text{ m}^2$
$1\text{-} 4 \text{ in.} \times 8 \text{ in. brick} = 200 \text{ cm}^2, 10{,}000 \text{ cm}^2 = 1 \text{ m}$
$10{,}000 \text{ cm}^2 \div 200 \text{ cm}^2 = 50$
50 bricks are needed per square meter
$50 \times 17.28 =$ 864
5% for waste = 43

Total bricks needed 907

Additional brick would need to be ordered to allow for border bricks if they are used in rowlock or soldier manner. Their quantity should be

TABLE 11–1 ESTIMATING TABLE FOR PAVERS INSTALLED IN MORTARLESS PAVING (DATA FROM BRICK INSTITUTE OF AMERICA, BUILDERS NOTES NO. 7)

Paver Face Dimensions (actual inches) w x l		*Paver Face Area (in sq in.)*	*Paver Units (per sq ft)*
4	8	32.0	4.5
3-3/4	8	30.0	4.8
3-5/8	7-5/8	27.6	5.2
3-7/8	8-1/4	32.0	4.5
3-7/8	7-3/4	30.0	4.8
3-3/4	7-1/2	28.2	5.1
3-3/4	7-3/4	29.1	5.0
3-5/8	11-5/8	42.1	3.4
3-5/8	8	29.0	5.0
3-5/8	11-3/4	41.6	3.4
3-9/16	8	28.3	5.1
3-1/2	7-3/4	27.1	5.3
3-1/2	7-1/2	26.3	5.5
3-3/8	7-1/2	25.3	5.7
4	4	16.0	9.0
6	6	36.0	4.0
7-5/8	7-3/8	58.1	2.5
7-3/4	7-3/4	60.1	2.4
8	8	64.0	2.3
8	16	128.0	1.1
12	12	144.0	1.0
16	16	256.0	0.6
6	6 Hexagon	31.2	4.6
8	8 Hexagon	55.4	2.6
12	12 Hexagon	124.7	1.2

Note: The above table does not include waste. Allow at least 5% for waste and breakage.

calculated by first determining the length of border in feet and inches or meters; then by multiplying the number of bricks per foot or meter, the total quantity needed is obtained. For example,

	200 ft of border bricks in soldier manner:
	1 brick spans 4 in. or 10 cm
Therefore:	3 per foot × 200 ft = 600 bricks
or:	10 per meter × 60 m = 600 bricks

PRACTICAL ASPECTS OF CONSTRUCTING A BRICK PATIO OR SIDEWALK

The several methods of construction discussed thus far are those using redwood 2 × 4s or brick pavers for border, or if a concrete base is used, no border may be used. Recall that the plans establish the widths and lengths as well as the heights needed, together with the proper slope. To interpret the various aspects used during the construction of patios or sidewalks, a patio with a grid of redwood 2 × 4s and a sidewalk with a concrete base will be used.

Patio with Redwood 2 x 4 Grid

Figure 11–11 shows various states of completion of the sample patio. Remember that these are the same conditions that are applicable in constructing mortarless sidewalks. At the beginning of the project, earth is removed within the area of patio to a depth approximating its finished height *less a 2 × 4's width* [3.5 in. (9 cm)]. Several easy methods can be used to keep track of the digging. Stakes can be driven to the outer edges of the patio. Then level and grade marks can be made on the stakes with a pencil once the line and line level are used to establish the proper height. A chalk line is usually stretched from stake to stake and tied on these marks. The depth of the excavation can easily be verified with a ruler from the line. Another method includes batter boards on each outer corner; again lines are used, along with a ruler, to verify the excavation depth. Regardless of the technique employed, the objective is to remove earth within the entire patio area to an approximate depth that makes construction of the grid easy and minimizes the need for excessive backfill.

Following the excavation, the 2 × 4 grid is installed. This should start with perimeter ones first; then intermediate pieces should be cut

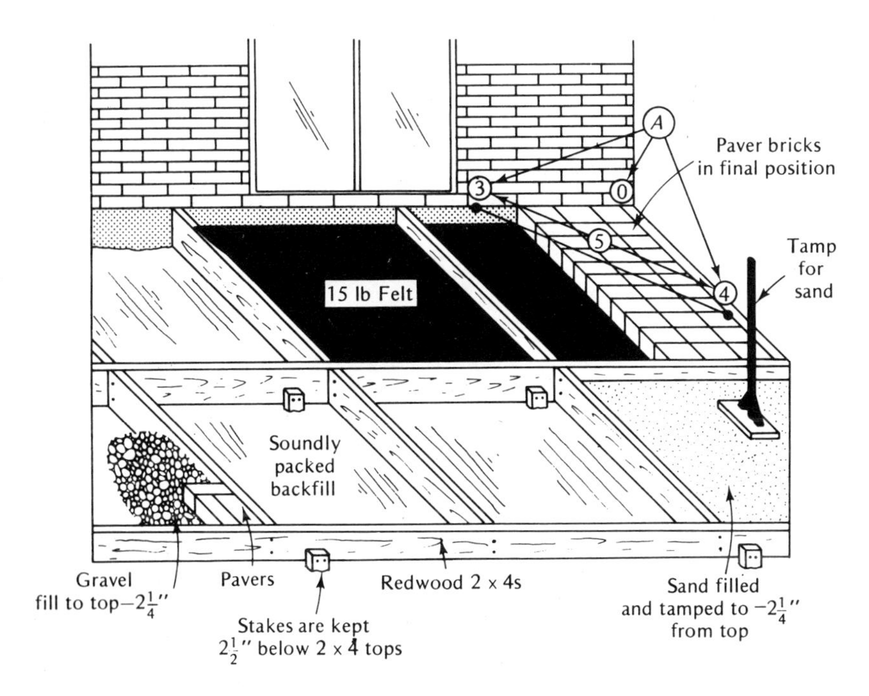

Figure 11–11 Patio with Grid and Mortarless Pavers

and fit into place. If a right angle is needed from the house to the patio grid, the 3–4–5 or 6–8–10 method should be used. This means that a point is measured 3 ft along the wall, 4 ft along the 2 × 4, and 5 ft from points measured (see Figure 11–11, callout A). Another method that can be used requires a steel tape measure. A diagonal length (corner to opposite corner) is measured and compared with the opposite diagonal. After several adjustments of form members, the lengths will be equalized, and the perimeter form may be considered squared.

At this time, stakes (1 × 2 × 12 in.) should be driven on either the inside or outside of the perimeter forms. Form height must be established with the slope included; then the stakes can be nailed to 2 × 4s with 6d or 7d galvanized common nails. See the note near the bottom of Figure 11–11, which states: "Stakes are kept 2½ in. below 2 × 4 tops." It is important that all stakes used within the form adhere to this requirement; however, those outside the form need not. The ones within the form must not interfere with the placement of brick pavers that are 2¼ to 2½ in. thick.

With perimeter 2 × 4s secured, intermediate 2 × 4s should be installed and nailed. These, too, may be staked so that additional support is provided. Toenailing is the preferred method, except when through-nailing with 16d common nails can be used effectively. It is absolutely essential that all top edges of 2 × 4s be kept flush (even). Their top edges are the guides from which all pavers will later be referenced.

Subgrading and Preparation

The next building step is to prepare subgrading for pavers. Recall that either gravel or sand may be used. If drainage is to be considered and used under the patio, all or a portion of its subgrade should be prepared with gravel. This may mean further excavation of subsoils to allow for placement of larger gravel and drainage tile. Under no circumstances should forms already installed be disturbed. The top layer of gravel should be of the pea gravel variety, and it should be brought to a height of 2¼ in. from the form's top after thorough tamping (see Figure 11–11, lower left).

If drainage is not a problem, sand should be used for fill. It should be misted with water and tamped to a finished level of 2¼ in. from the form's top. Then as Figure 11–11 shows, pieces of 15-lb felt paper or polyethylene are laid over the sand to control water runoff and to discourage plant and grass growth. When this is done, the pavers may be installed.

Installing Pavers with or Without Cement

Any of the designs shown in Figure 11–5 can be used during paver installation. In a mortarless patio, all pavers are placed close to one another with spacings up to ¼ in.; therefore, the best sizes are either square or those whose lengths are twice their width.

A straightedge and mallet are used to align the paver bricks within the 2 × 4 forms. Gentle taps with the mallet depress the paver into the sand, and the straightedge verifies alignment (levelness). If felt paper is used, caution should be exercised to avoid tearing it. All work should start at the perimeter so that no disturbance of either packed sand or tamped gravel occurs. Using this sequence of installation allows the bricklayer to work on bricks already installed.

After all pavers are installed, cement may be used right from the bag as fill between pavers, or it may be mixed with fine sand. If used, it should be broomed into the joints with a backward-and-forward motion. Then all excess cement or cement and sand must be swept away. Finally, the surface is hosed with a gentle spray of water. Settling usually occurs, so the process must be repeated several times.

If wider than minimum joints are the rule rather than the exception, a mixture of water and mortar consisting of fine washed sand and cement (3 parts sand to 1 part cement) should be used as grout. Since these joints may be rather wide, they should be struck with a concave tool. If either method employing cement is used, some residue is likely to remain on brick surfaces. Therefore, the entire surface must be washed with an acid solution.

If cement is not desired, fine sand should be spread and broomed into all joints, then sprayed with water to cause it to settle. This process may be repeated until the sand is firmly packed. This method is not as good as the one using cement because grass and weeds may easily root in the sand.

Sidewalk Constructed with a Concrete or Mortar Base and Brick Pavers

Figure 11–10 detailed three elevation plans that show how a sidewalk may be constructed when either concrete or mortar is used as its base. Each drawing depicts requirements of construction that differ slightly. The selection of one of the three should be decided upon based on the variables listed at the beginning of the chapter. Some of these were subsoil conditions, types of soil, drainage needs, and art form desired.

If a construction method using elevation a or b in Figure 11–10 is selected, no border bricks are needed. The concrete base provides the stability for the pavers that is absent in the mortarless construction technique. However, if a construction method using elevation c of Figure 11–10 is elected, border bricks must first be anchored in concrete. This method involves considerably more work, materials, and time, but there may be sound reasons for its use.

Sidewalks Without Border

Figure 11–12 is a modification of Figure 11–10a and b to provide a pictorial aid in describing the construction tasks involved. In Figure 11–12a the forms for the sidewalk are established at either 2 in. or 2¾ in. below finished grade level. This requirement can also be stated as *grade minus 2 in.* or *grade minus 2¾ in.*—grade being, of course, the top surface of each paver.

At this time a decision must be made about the method of instal-

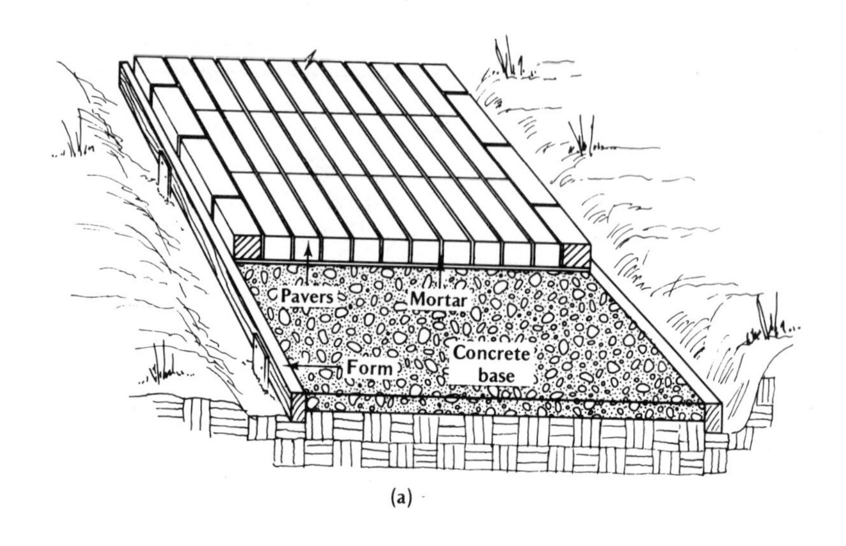

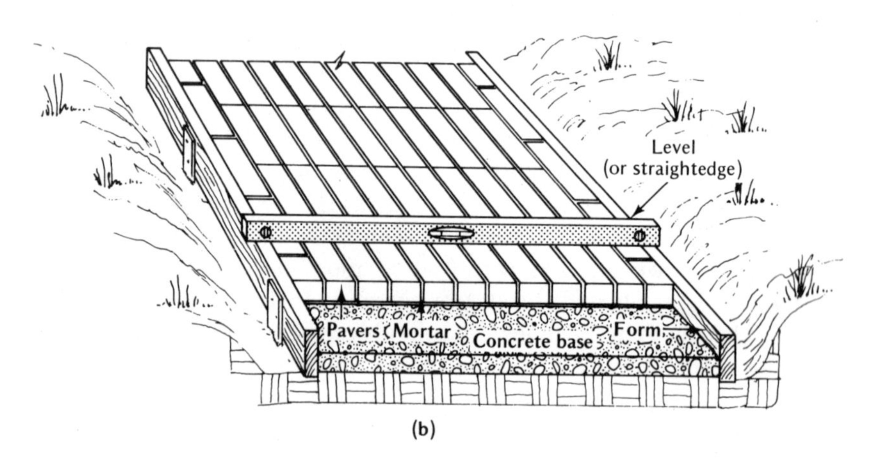

Figure 11–12 Sidewalk with Concrete Base and No Border Bricks

ling the brick pavers. The decision affects the height of the forms, as shown in Figure 11–12a. If each paver is to be anchored directly into the poured base while the base is in its plastic state, the forms must be set at grade minus 2 in. Each brick paver is set into the base at least ¼ in. If on the other hand, the pavers are to be mortared in place after the base has set, then the forms must be set at grade minus 2¾ in. This allows for a 2¼-in. brick plus ½-in. mortar bed for the brick.

In the event that the brick pavers are embedded in the concrete base, mortar is inserted between joints with a trowel. Then the joints are struck and made concave. Where the bricks are mortared over the set concrete base, mortar joints should be made by buttering sides and ends as each brick is laid in place.

There may be a problem of maintaining a uniform height or grade level. Some type of depth or height gauge may be made or a mason's line may be strung across or along the sidewalk to establish the finished grade level. If a gauge is made, it should be easy to handle and readily available. It could simply be a piece of 2 × 4 about 6 in. long, ripped to a width equaling the distance from base to finish grade. If a line is used, an outer course of bricks should be installed; then the level should be used from then on. Or both outer courses can be laid using the line, and from then on the level is used to establish and verify heights of all the pavers between the outer ones.

Figure 11–12b illustrates a construction technique slightly different from the one described previously. When using this technique, all forms are set to the *finished grade level.* Concrete is placed into the form and screeded to a level of either grade minus 2 in. or grade minus 2¾ in.

The pavers are then installed using either the embedded method or mortar method. The difference is that the forms are set at grade level and all brick pavers are established for grade from these form members. In addition, mortar placed between pavers is restricted from lateral motion (oozing) by the forms, and alignment of pavers along each edge of the sidewalk is perfect.

Sidewalk with Border

The pictorial (Figure 11–13) of the elevation drawing (Figure 11–10c) shows the construction techniques needed as:

1. Setting footings to anchor soldier-coursed border bricks.
2. Backfilling between footings and tamping.
3. Concrete base installation and screeding.
4. Mortaring pavers between and flush with border bricks.

The position, grade, and width of the sidewalk must be defined from the plot plan. A mason's line and grade stakes must be used for the purpose of establishing where each border of bricks is to be in-

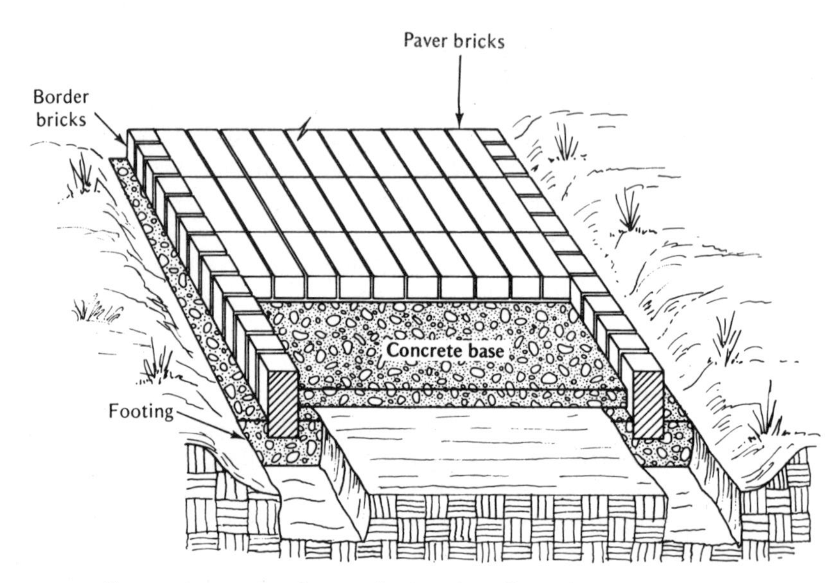

Figure 11–13 Border-Bricked Sidewalk with Concrete Base

stalled. The ground must be removed to allow for the placement of the footings for the border bricks. Forms are not needed as a rule, since this earth probably will suffice as form walls, and the concrete is to be covered later. Once the concrete has been poured and screeded approximately even, a row of border bricks is set deeply into the concrete. Each brick must be lined very carefully with the mason's line for both height and horizontal position.

Backfill, tamping of the backfill, and placement of the concrete base are tasks that follow installation of the border bricks. However, these tasks must *not* be performed until the borders are set (24 hr or longer). Backfill of dirt, or gravel and dirt, or just gravel should be brought to a level equal to grade minus 2 in. or 2¾ in. *plus* the thickness of the concrete base (approximately 2 in. more).

The concrete is installed next, tamped thoroughly, and screeded to the level needed for installation of the pavers. Following the placement of the concrete, the pavers are mortared and the mortar joints are finished.

Since mortar is used, it is impossible to keep mortar from brick surfaces. Washing with an acid solution followed by a clean water rinse is recommended shortly after the mortar has set and cured.

INSPECTION OF BRICK PATIO OR SIDEWALK

Inspection	*Satisfactory/N.A.*	*Unsatisfactory*
Was the site for patio or sidewalk properly defined on the plot plan?	______	______
Are all dimensions, including grade levels, detailed?	______	______
Are there any drainage problems?	______	______
Were provisions made for the correction of drainage problems?	______	______
Was a design pattern selected that contributes to development of the feelings intended?	______	______
Were proper paver sizes ordered and delivered?	______	______
Were border bricks embedded in concrete?	______	______
Was proper subgrade sand or gravel used?	______	______
Was subgrade fill tamped thoroughly?	______	______
Was 15-lb felt cut and installed?	______	______
Were pavers placed as close as possible during installation?	______	______
Was concrete or concrete and sand installed properly and dampened?	______	______
Was sand washed well into the joints between pavers?	______	______

QUESTIONS

1. What is a brick paver?
2. What three important aspects of a patio or sidewalk need to be learned from plans and specifications?
3. What is a proper slope for a patio?
4. Which blueprint drawing shows the slope and cross-sectional details of a patio or sidewalk?
5. Name two important reasons for sloping a patio.
6. List three ways in which mortarless pavers may be kept in place.
7. What is the best solution if the soil below a patio contains large quantities of clay?
8. Why would it be necessary to include gravel below the pavers?
9. What material is used below the sand base and paver that inhibits plant growth?
10. What grade should paver bricks be classified if the best weather wear is expected?
11. Calculate the number of pavers (4 in. × 8 in.) needed for a sidewalk 3 ft wide by 36 ft long.
12. Why must stakes that are nailed to 2 × 4 grids be kept 2½ in. below grid top?
13. Why are pavers laid from near to far?

Chapter 12

Block Walls and Foundations

Block a concrete masonry unit made with fine aggregate and cement shaped in a mold. Any of a variety of shaped light or standard-weight masonry units.
Clinker a small portion of mortar that protrudes through a wire mesh and dries, forming a bond between mortar and screen.
Crawl space 18 in. or more of space between the ground and underside of floor members.
Finished floor height a line or an elevation-plan indication of the absolute height of the first floor in a building.
Furrowing striking a V-shaped trough in a bed of mortar.
Girder a wooden or metal beam that spans a long, open area and rests on pilasters or columns.
Grout a loose mixture of portland cement, aggregate, and water which is poured into hollow cells or joints in masonry.
Joists a wooden horizontal member used in floor or ceiling construction (e.g., floor joists, ceiling joists).
Ledger a wooden member that is used as a ledge to support other materials, such as joists.
Lintel a beam placed over an opening in a wall.
Mortar a plastic mixture of cementing materials, water, and aggregate used to bind masonry units.
Pilaster a wall section that is reinforced to serve as a column and may project in or out from the wall's line.
Rafter a wooden member used in constructing a roof.
Rod a steel reinforcing bar.
Running bond the same as common bond, with horizontal joints continuous but vertical joints offset or in line.
Truss a wooden frame used in roof construction; may also be made of metal.

OBJECTIVES—INTRODUCTION

The block used in building is a relatively new product compared to brick. It has many desirable features, and the large variety of shapes now available makes it extremely versatile. Some of its desirable features include cavities that create dead-air (insulating) pockets; size, which permits construction of 4-, 6-, 8-, or 12-in. thick walls with a single unit; bulk, which allows erection of a large wall relatively quickly; and design, which allows for easy reinforcement. There are many, many shapes and sizes of blocks; some are designed for architectural beauty while others are designed for construction needs, such as the J block used in foundation-to-floor work.

This chapter examines the blueprint and specifications commonly used with block construction and discusses several requirements for sound construction techniques. This leads to identification of the two objectives for this chapter: *to be able to formulate a plan for block-wall construction,* and *to understand the elements of block-wall construction as related to support and stability requirements.* These two objectives imply a need to define and understand such parts of the wall as footing dimensions, reinforcement criteria, grouting, methods used to create walls, pilasters, lintels for openings, and block-laying techniques. So that a comprehensive check can be quickly obtained for block-wall construction, the chapter closes with an inspection checklist.

READING A BLUEPRINT FOR A BLOCK WALL AND FOUNDATION

This section discusses and illustrates various segments of blueprints, including blocks set onto or into footings, J blocks used in slab construction, common blocks used in residential home walls or in retaining walls and their reinforcement and grouting needs, and the construction of lintels over door and window openings using the lintel block.

Foundations of Masonry for Wood-Frame Buildings

The elevation plans and section or detail plans provide valuable information for constructing the foundation of a building. Blocks are mortared onto concrete footings and raised to floor height. A wooden frame is used above the foundation. Several methods of constructing the foundation are illustrated below.

1. The full basement; wooden floor joist. Figure 12–1 shows a section elevation drawing of a full basement block wall, its

footing, and the floor joists and stud wall. The footing should be poured 8 in. (20 cm) thick by 16 in. (40 cm) wide for one- and two-story houses. This assures an adequate base for the combined weight and force of blocks used for the basement wall, plus the weight of the sheathing and roof of the house.

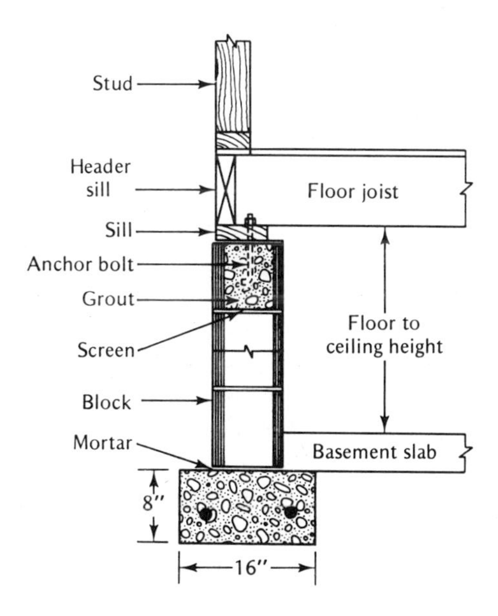

Figure 12–1 Section—Footing and Wall

The construction detailed in Figure 12–1 includes mortaring the first course of block onto the previously poured and set footing. From there on, blocks are laid according to specifications, which could be: "Lay blocks using running bond with ⅜-in. mortar joints; strike joints using the concave technique." The last or top course of block is to be filled with grout, but if provisions are not taken, grout will fall to the footing level. One common method is to lay a strip of wire lath onto the course just below the final course, as shown in Figure 12–2. This metal, which is cut approximately 5 in. wide, is centered on the blocks. The final course of blocks is laid. Then grout is poured to the level of the block. This makes a continuous bond. While the grout is still in a plastic state, anchor bolts are installed.

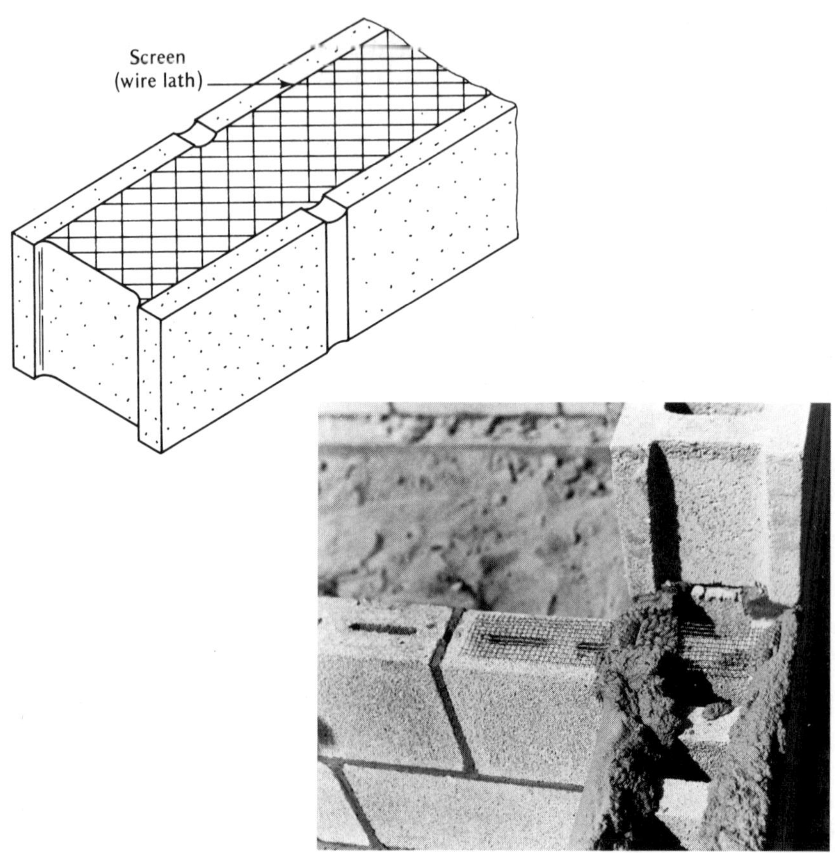

Figure 12–2 Wire Mesh to Support Plastic Grout *(Photo Courtesy of Portland Cement Association)*

2. The crawl-space foundation. The description just completed could easily be used for a wood-frame building with a crawl space. However, if the crawl space is not very high, a single U block could be used. Figure 12–3 shows the sectional elevation for this method of construction. Notice that the block is set into the footing while the footing concrete is still in a plastic state. This method eliminates the need for mortaring the block to footing. However, the task is not as simple as it appears. Vertical joints between blocks must be mortared either as they are being laid or after the footing has set. Then, too, all blocks must be *in line* and accurately aligned for height to within 1/4 in./10 ft of run.

Figure 12–3 U Block and Crawl Space

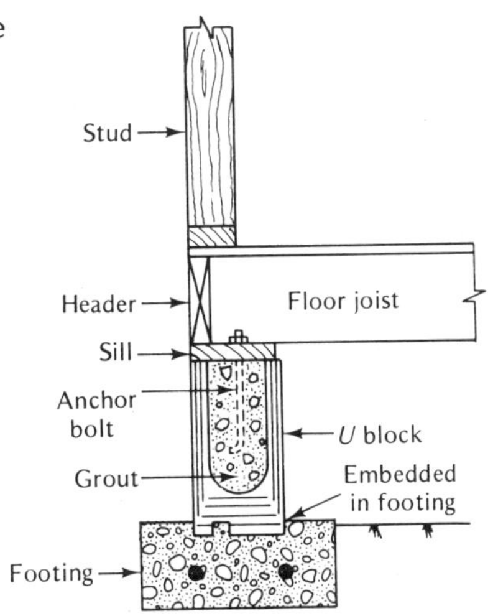

3. J-block and concrete floor. One specially designed block frequently used in foundation construction where the main floor of the residence is poured concrete is the J block. As Figure 12–4 shows, this block is molded with one flange lower than the other, which is a modified U block. The taller side (flange) is customarily installed outward from the slab and its top surface or edge is set to the finished floor height. Earth is backfilled and compacted to the lip of the inner, lower flange. Before concrete is poured, all plumbing is roughed in and plastic sheet goods are laid over the fill. Notice that the concrete is allowed to fill into the block. This develops a unified structure. While the concrete is plastic, anchor bolts are installed.

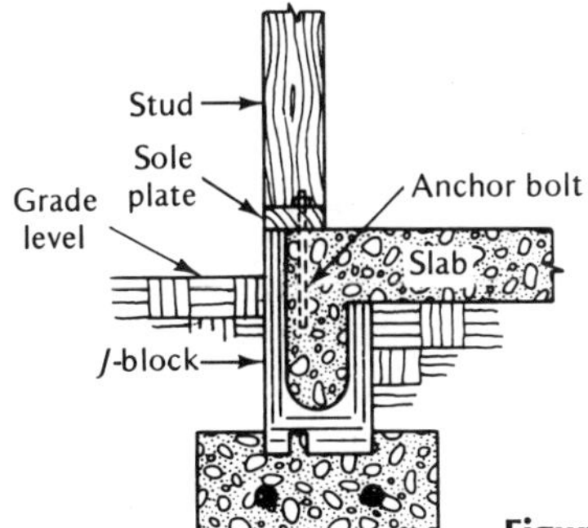

Figure 12–4 J-Block Foundation and Slab

Figure 12–5 J Block, Slab, and Block Walls

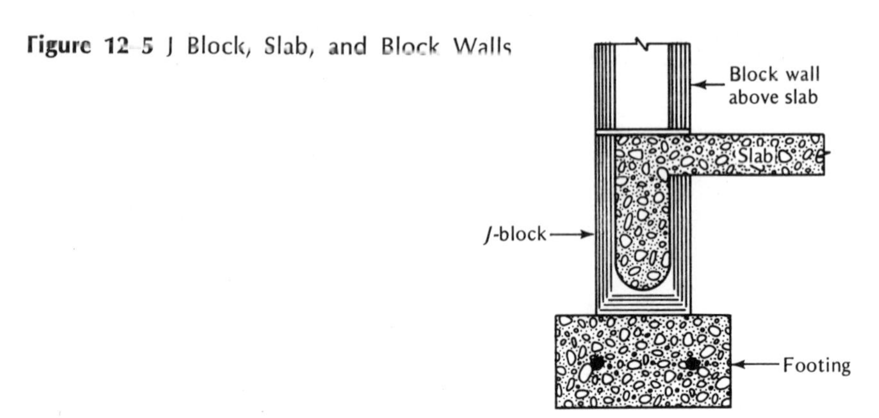

Foundations and Walls Using Block Masonry

Where a slab floor is used in conjunction with block masonry foundations, the walls of block above floor level are laid on top of the poured floor, as shown in Figure 12–5. In contrast to the blocks used in the foundation, which are filled with grout, the wall blocks are not uniformly filled with grout. Only those hollows that are to have steel reinforcing rods are filled.

Several other methods of constructing the block wall and uniting both foundation and floor(s) are available to the builder. The elevation sections shown in Figure 12–6 illustrate these. In Figure 12–6a, a continuous alignment of blocks is maintained and the concrete slab is united to the foundation wall with horizontal rods inserted into the block wall. Where wooden floor joists are to be used, a ledger of nominal-3-in. stock is bolted to the foundation as is a header above the ledger, as shown in Figure 12–6b. This means that these anchor bolts must be installed while the masonry units are being laid up. Standard practice is to set the spacing of bolts at 48 in. on-center. However, local conditions, as well as predicted live and dead loads, may require closer spacing. Finally, Figure 12–6 shows that a foundation of 12-in. blocks may be laid up to floor sill height; then the remainder of the wall is laid up with 8-in. blocks. This technique develops a masonry ledge for support of the floor joists. Even with this arrangement, though, anchor bolts are needed to secure the floor joist header to the foundation.

Lintel over Window and Door

Block manufacturers have designed and made available several types of blocks that are used specifically for lintel construction. The one in

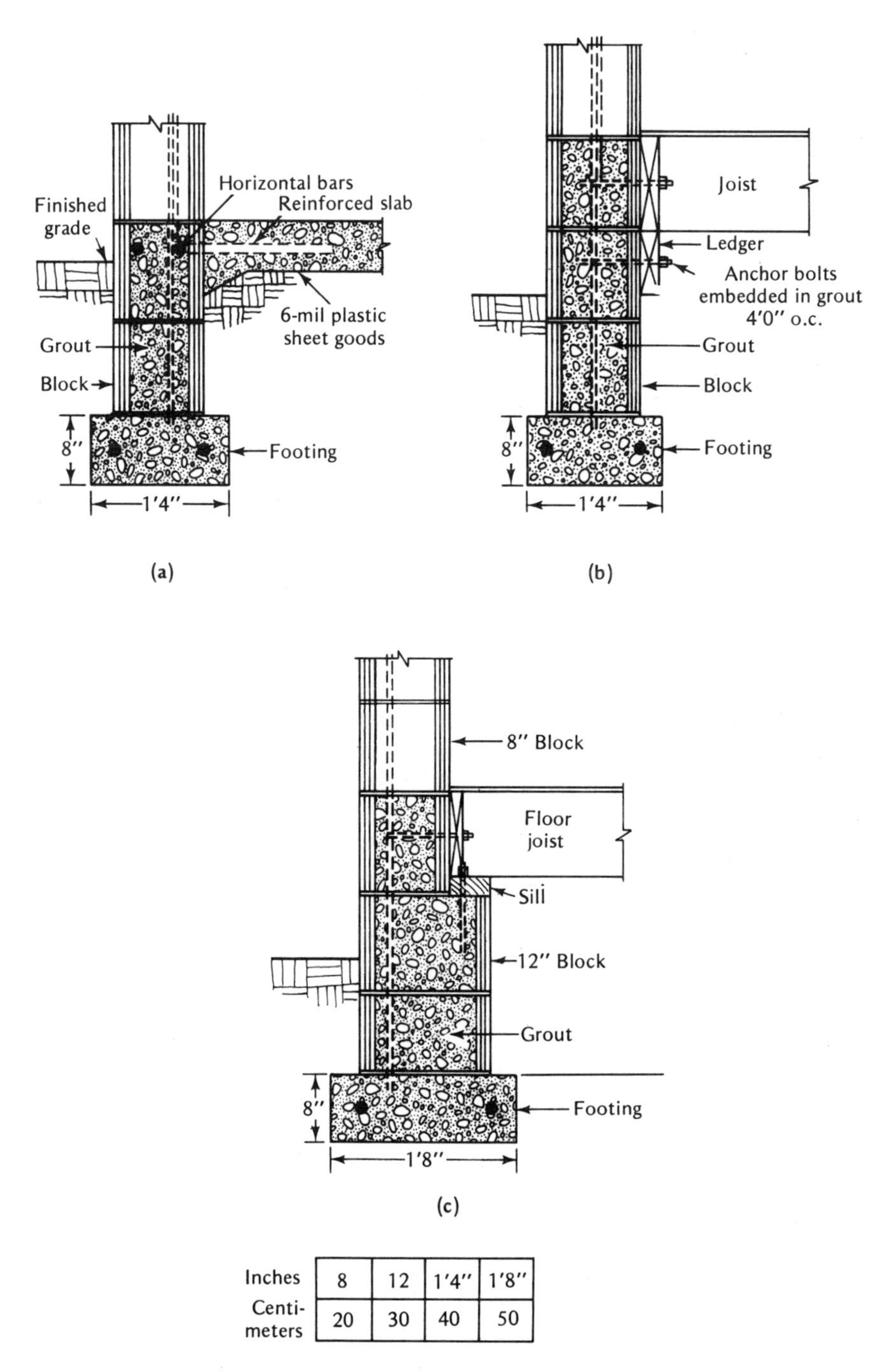

Inches	8	12	1′4″	1′8″
Centi-meters	20	30	40	50

Figure 12–6 Three Techniques of Block Foundation and Wall Construction

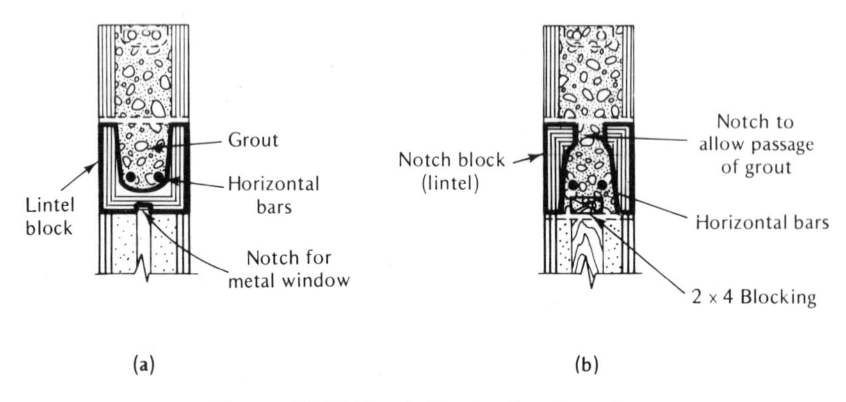

Figure 12–7 Lintel Blocks for Openings

Figure 12–7a is like a U block, but for lintels the height of the block is reduced. This block has several advantages: (1) the cavity forms a natural pocket for containing the plastic concrete and steel rods, and (2) the notch in the base of the block is used where metal window-sash is called for. The use of this block eliminates some of the need for forming requirements during the construction of lintels.

The notch block shown in Figure 12–7b provides an additional advantage over the first one described; it allows for the inclusion of backing materials, such as the 2 × 4 shown. This block also confines the plastic concrete used as grout, thereby eliminating the need for forming. It does appear, however, that some forming, at least a soffit, must be used for the notch block.

All lintels except precast types require a soffit for support of the lintel block. The soffit is usually made from a nominal 2 × 8 member for an 8-in. wall (2 × 4 for 4-in. wall, 2 × 6 for a 6-in. wall, and 2 × 12 for 12-in. wall). As shown in Figure 12–8, it is adequately supported by perpendicular members of 2 × 4s or 4 × 4s. The expanded view A illustrates the need for installing the soffit ⅜ in. above the course of block. Blocks either side of the opening are bonded with mortar; those over the opening are not. Therefore, adjustment must be made to maintain horizontal lines. If ½-in. joints are used for mortar, the soffit would be raised ½ in. above block level.

It is not always necessary to use lintel blocks when making a lintel. A *bond beam block* (Figure 12–9) can be used. The steel rods are placed along the recessed cavity, and grout is filled in to complete the lintel.

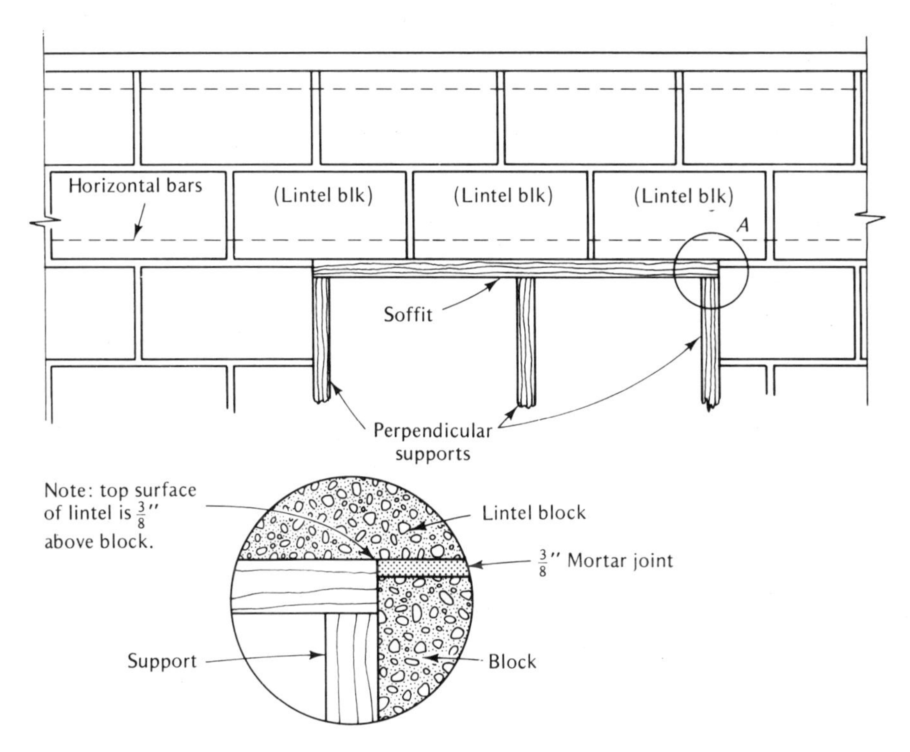

Figure 12–8 Lintel Support During Construction

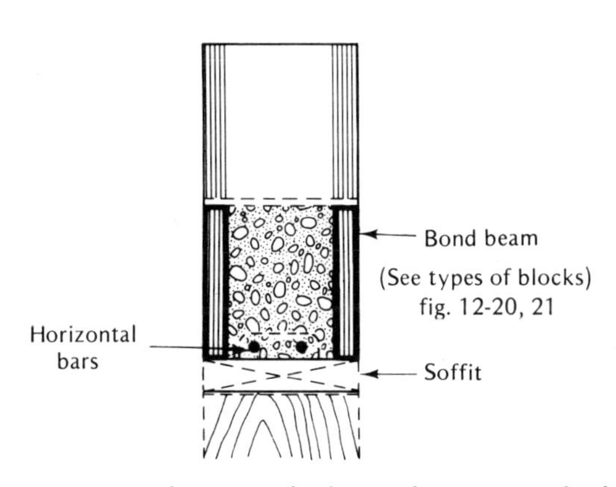

Figure 12–9 Bond Beam Block Used as a Lintel Block

Capping the Wall

In most residential and some commercial block-wall construction, window heights are established as 12 in. (30 cm) or 16 in. (40 cm) from the ceiling. This allows for a continuous bond to be included around the building at the ceiling level (top of the wall). Figure 12–8 shows the continuous bars at the lintel level. The implication here is that bond beam blocks are used on the wall. Bond beam blocks are also used on the final or top course of block.

Before the first course of bond beam blocks is set, wire mesh should be laid across the course, as shown in Figure 12–2. After that, the horizontal bars are set and blocks are laid. The course is filled with grout. Next, the top course of blocks is laid up, but this time the recessed area in the bond beam may be placed right side up. After laying the course, grout and reinforcing bars are installed. The grout should be struck flush with the top of the block, and then anchor bolts should be set in place in preparation for the sill.

The elevation section for the top two courses looks as shown in Figure 12–10. The sill is usually installed flush with the exterior surface of the block and held in place with the anchor bolts. The dashed outline illustrates how the ceiling joist and rafter appear when in place on top of the sill.

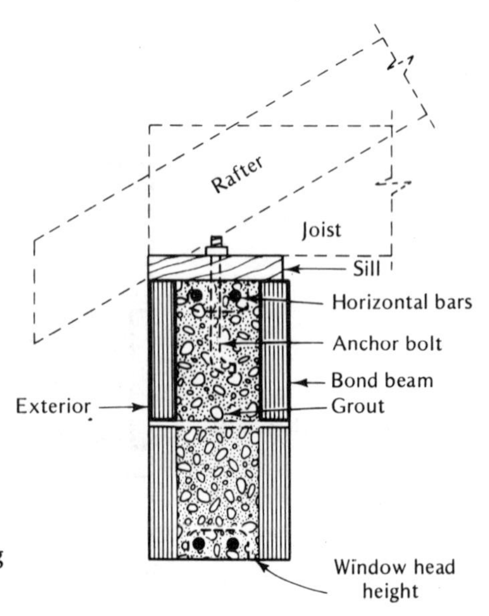

Figure 12-10 Section—Bonding the Final Two Courses

Vertical Reinforcement Within the Wall

Thus far in this chapter sectional details have been explained so that their importance can be recognized. The progression of explanations follows the usual installation sequence. Foundation tasks are done before floor and lintel tasks, and bond beam uses at the top of the wall are left until last. However, throughout this progression, vertical bars are installed at strategic points.

Figure 12–11 shows a full length of wall with the following requirements that affect steel-rod placement:

- **a.** Corners.
- **b.** Window.
- **c.** Door.
- **d.** Steel-rod separation.
- **e.** Interior wall.
- **f.** Steel rods along side openings.

The drawing shows that bars are separated 4 ft 0 in. on-center except the spaces reserved for doors and windows. Wherever a door or window opening is made, a bar is installed in the hollow of the block each side of the opening. Grout is filled in to hold the steel in position and to unify the assembly.

The first question that one raises about using vertical bars for reinforcement is: How are blocks positioned over a rod that may stand 8 ft high? As a rule, the rod must be tied to the one protruding from

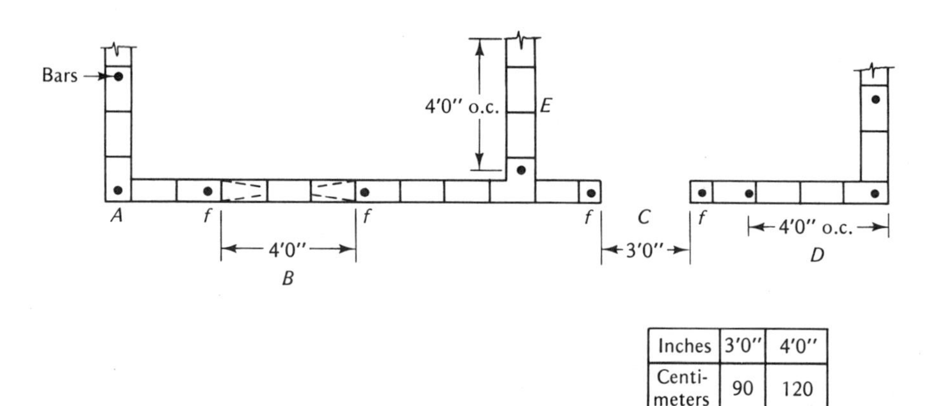

Inches	3'0"	4'0"
Centi-meters	90	120

Figure 12–11 Vertical Rod Placement

the foundation or footing. The answer is to use shorter lengths of rod and wire them together. But this raises a question: What is the required overlap needed? Overlap of the rods should be set at a minimum of *30 times rod diameter*. For several examples, look at the following chart.

Bar No.	*Bar Diameter (in.)*	*Overlap (30 × Diam)*
3	⅜ (0.375)	11¼ in. 28.1 cm
4	½ (0.500)	15 in. 37.5 cm
5	⅝ (0.625)	18¾ in. 46.9 cm
6	¾ (0.750)	22½ in. 56.3 cm
7	⅞ (0.875)	26¼ in. 65.6 cm

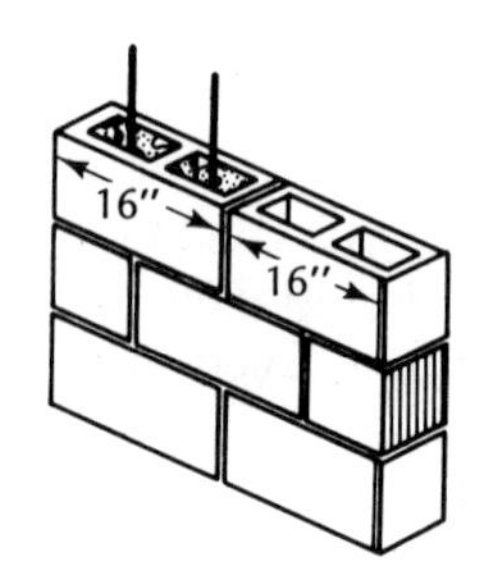

Pilaster from common block and reinforcement

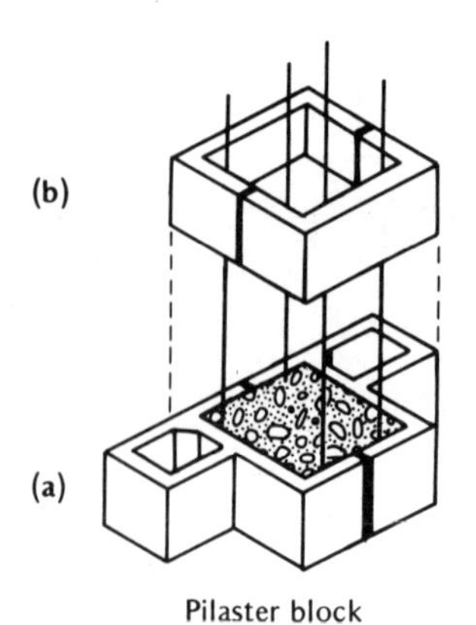

Pilaster block

Figure 12–12 Pilaster of Block

Pilaster Made from Block, Steel, and Grout

The plans of a house often call for a pilaster. Several poured-concrete construction methods were shown in Chapter 2. Blocks can also be used to make pilasters. In one method the common block is used, as shown in Figure 12–12. Notice that the pilaster shown actually is common block with steel reinforcing rods and grout. If this method is used, no projection from the wall either inside or outside is made.

However, sometimes it is not sufficient just to strengthen a segment of wall to support a truss or girder. Frequently, the girder or truss must be installed within the wall. In this method two blocks are especially designed for the purpose. The block shown as A in Figure 12–12 allows for continuation of a common 8-in. (20-cm) wall and integrates the pilaster. The block labeled B is designed for the alternate course. Once mortared in place, the cavity, with four reinforcement bars installed in the corners, is filled with grout.

Retaining Wall; Its Cap and Reinforcement

Building a reinforced retaining wall from blocks requires most of the essentials of all block walls. However, blocks need to be laid on a wider-than-usual footing, with most of the footing toward the backfilled area, as the sectional drawing (Figure 12–13a) shows. The blocks are usually laid in a running-bond pattern because of the added strength that is obtained. The vertical bars used in reinforcing the wall are installed at a maximum 4 ft (120 cm) on-center, and all cavities that have bars are filled with grout. The top course of block could very well be made from bond block so that two horizontal bars may be installed continuously through the full length of the wall. If this is done, wire mesh must be laid on top of the course below the bond beam blocks to restrict grout from dropping into lower cavities.

Since the wall is to be exposed continuously and be subject to all weather conditions, its cap must shed water. Several caps are shown in Figure 12–13, together with the necessary forms. Figure 12–13b shows a sloping cap or shed slope, with both edges flush with the wall surfaces. Usually, 1 × 6 materials are sufficiently strong to contain the plastic concrete cap provided that the braces are placed close together (30 in. o.c.). A more decorative cap can be installed on a retaining wall as shown in Figure 12–13c. This one includes an overhang of 1½ in. (4 cm). It is made by using a 2 × 4 to create the overhang and 1 × 6 stock and braces.

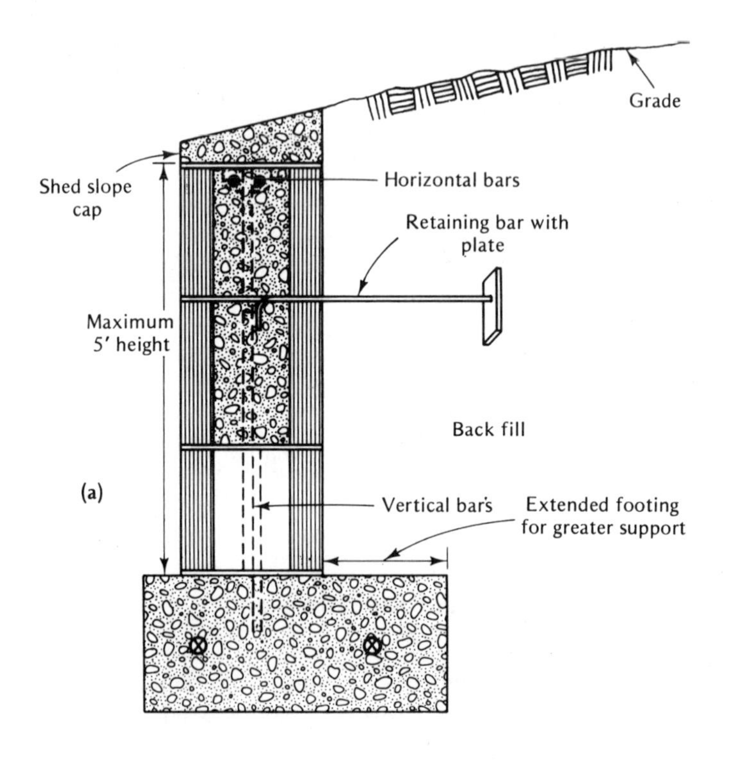

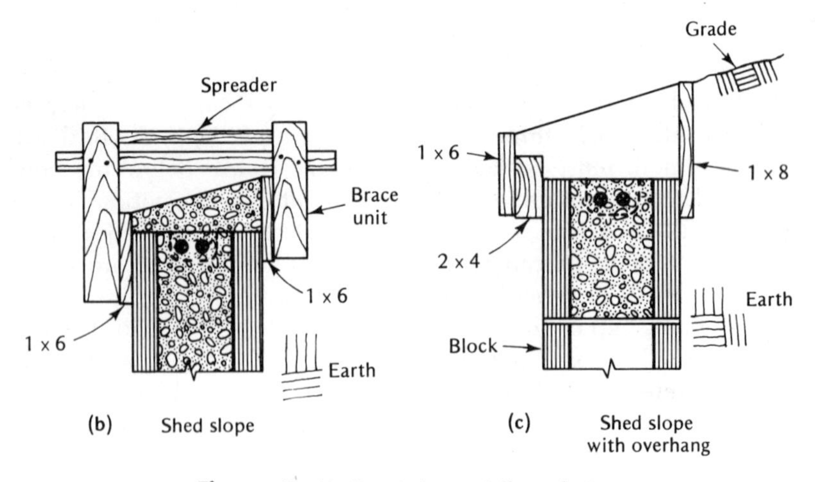

Figure 12–13 Retaining Wall and Cap

The amount or degree of slope should be judged according to the grade behind the wall. If the grade is shallow, the slope of the cap should be shallow. If the grade is steep, the slope of the cap should be steep. A single pour of concrete can be used to fill the bond beam and cap form if small course aggregate (pea gravel) is used.

Although the wall and footing are strong and can sustain lateral pressure such as that of the earth, retaining bars with welded plates can be installed. These should be installed periodically along the wall and may also be installed at varying heights (see Figure 12–13a). Their most useful positions are from one third of the way up the wall's height to one fourth of the way down from the top, and they should extend into the earth-filled area 3 to 5 ft. Grout must be used in the cavity of the block to secure the end of the bar.

BLOCK LAYING (STANDARD METHOD)

Block laying is the translation of all the principles used in planning a masonry building. Regardless of the size of block used, laying blocks follows basic techniques. Historically, block-laying techniques were derived from bricklaying techniques. At first blocks were made in very few sizes and shapes, but now a large variety of sizes and shapes is available. Many of those shapes are shown in the elevation plans in this chapter.

Generally, the block is laid in the following sequence:

1. *First-course layout.* Decide the layout sequence of the first course, which can be done in one of two ways. A block layout plan such as the one shown in Figure 12–11 may be drawn. This plan would indicate where full-length, three-quarter, one-half, and one-quarter blocks would be installed. It also could note where end or corner blocks are to be used. The second method is done on the job site. Blocks are laid dry along the footing with spacing for mortar joints and doors (if required). This method provides a visual approximation of the block positions and is generally used by apprentice and amateur masons.
2. *Prepare mortar.* Prepare a batch of mortar consisting of mortar mix and sand or portland cement, hydrated lime, and sand by volume. If masonry cement (mortar mix) is used, sand by volume should be added according to directions on the bag. If, on the other hand, portland cement is used, the proportions should be 1 part portland cement, ¼ part lime, and 3¾ parts sand.

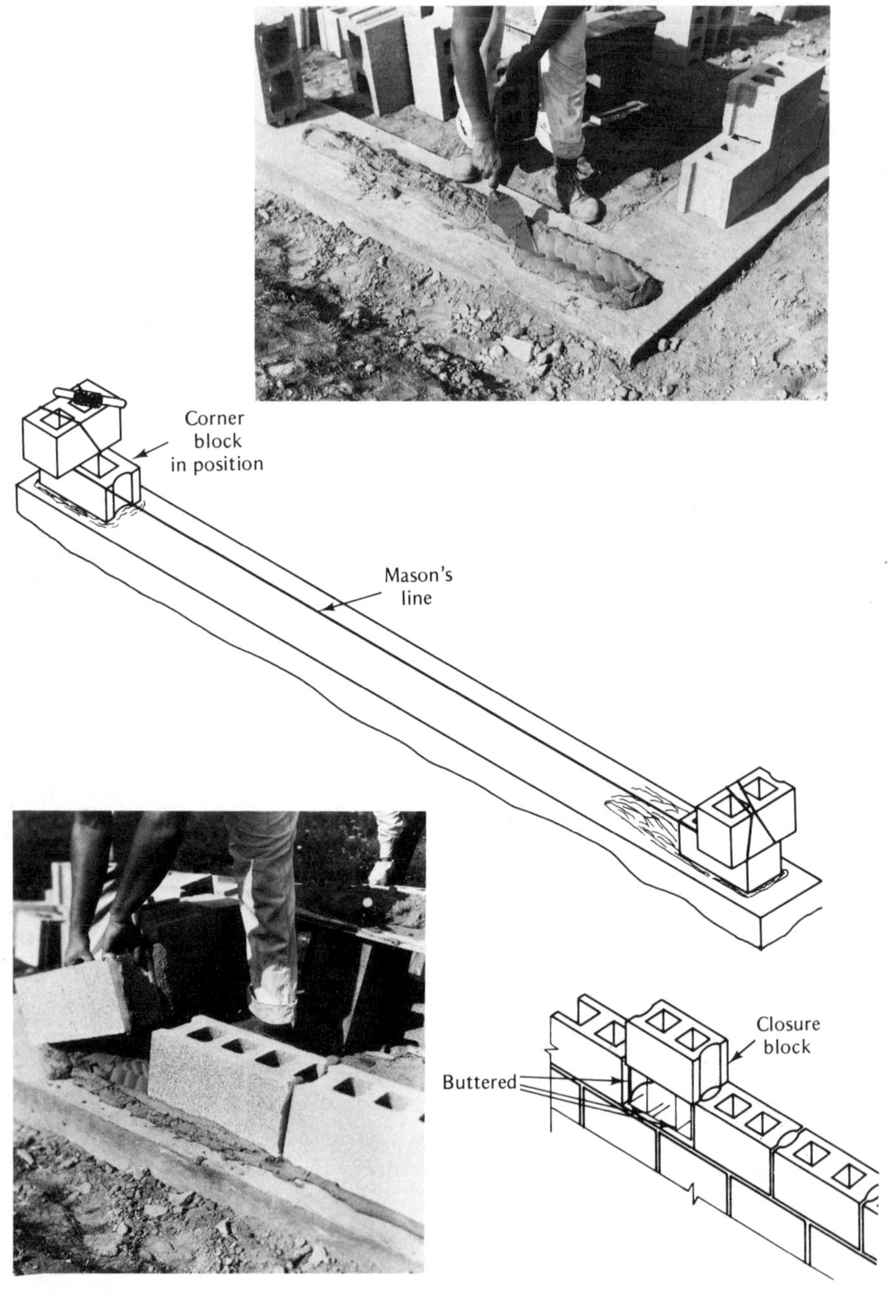

Figure 12–14 First-Course Design (*Photos Courtesy of Portland Cement Association)*

3. *Mortar the first course of block to the footing.* All or some of the blocks positioned dry on the footing (if done) should be moved off the footing and in their place a generous amount of mortar should be laid. The trowel should be used to spread and furrow the mortar (Figure 12–14a).

a. A corner block is laid first. It must be aligned true and plumb with wall lines previously established on batter boards.

b. Next, a corner block is laid on an opposite corner in like manner; between the two a mason's line is stretched, usually along the outer top edge. This is held in place with another block (Figure 12–14b). The line is used to establish alignment and uniform height.

c. Intermediate blocks are laid between corners by buttering the end projections of each block (Figure 12–14c). As each is laid, it is aligned with the mason's line, plumbed, and leveled.

d. The final block in the course is a closure block, and it is laid in a bed of mortar that includes mortar on the ends of both adjoining blocks (Figure 12–14d).

e. As each block is laid, the excess mortar is lifted away by using the trowel in a flat scraping and lifting movement.

4. *Raise the corners.* Raising the corners three or four courses usually follows the completion of the first course. This task is particularly important for several reasons. The corner must be absolutely plumb, since it establishes the line of the building. It also must be done properly so that vertical alignment of alternate courses on running bond or stack bond joints are perpendicular. It also must establish the height of each course. If, for example, one mason uses ½-in. joints on one

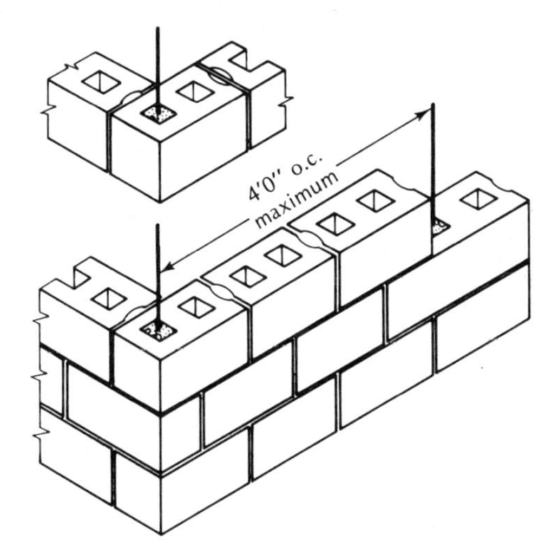

Figure 12–15 Corner Design

corner while another uses 3/8-in. joints on an opposite corner the horizontal accuracy for level is lost. Therefore, story poles similar to the one shown in Appendix A are used. Figure 12–15 shows a typical corner of a wall made from 8-in. blocks. Other patterns are available for 4-, 6-, 6- to 8-, and 12-in. walls. (See the Masonry Institute of America or Portland Cement Association entries in the References at the back of the book.)

5. *Door openings.* If an opening for a door must be made into the wall, its blocks either side of the opening should be treated as a corner and laid up several courses (Figure 12–16a). This method is especially appropriate where a doorjamb is not positioned prior to the laying of block.

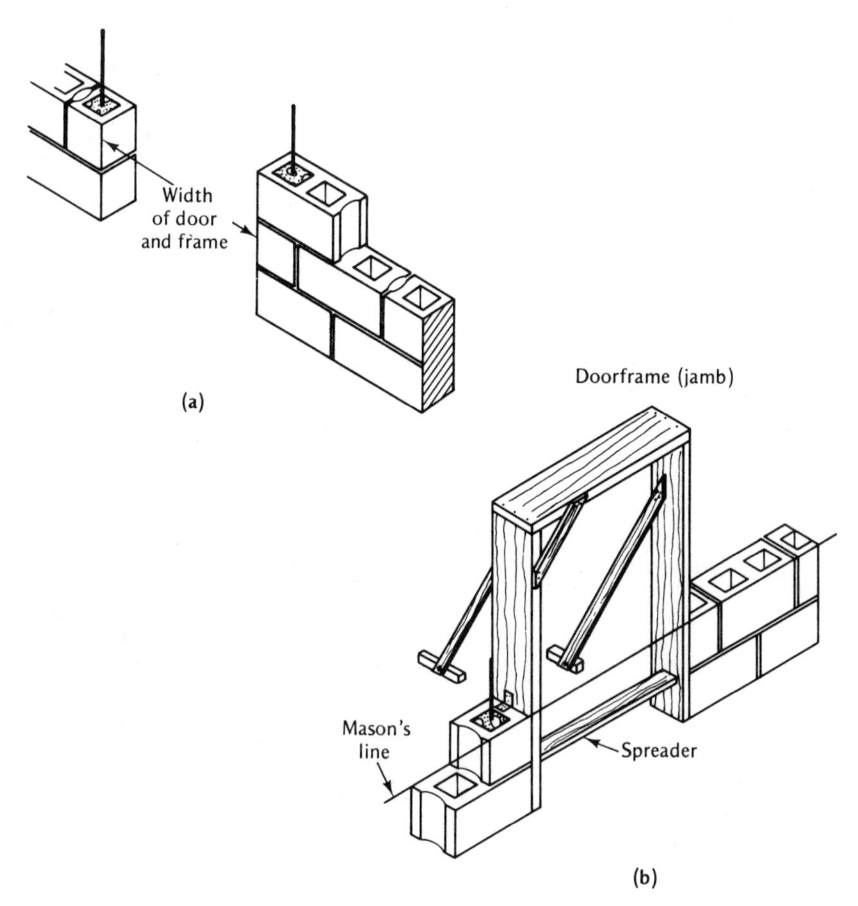

Figure 12–16 Door Openings

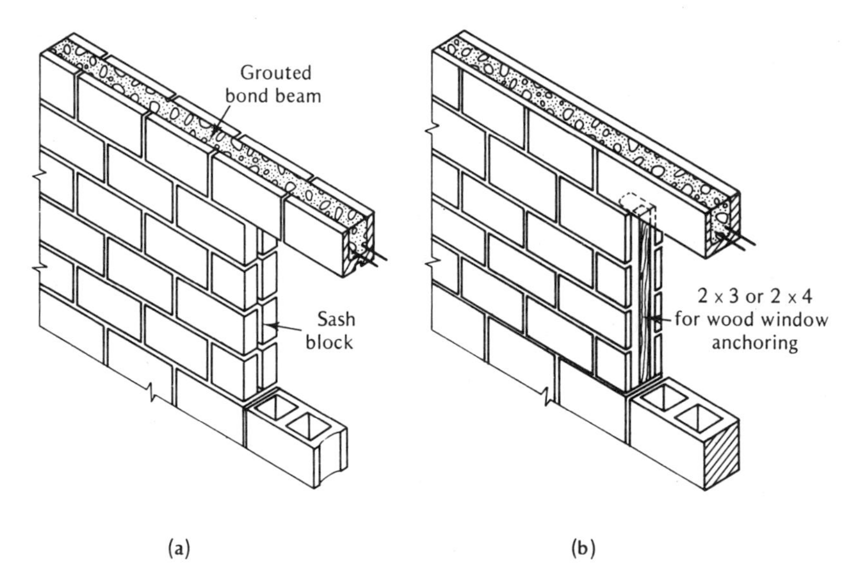

Figure 12–17 Window Openings

However, if the carpenters pre-position either the doorjamb or a complete door-frame unit prior to block laying, there is little need to treat the adjoining blocks as corners. A mason's line can be strung along the inner or outer edge where the jamb or frame does not interfere and blocks can be cut to fit as they are laid (Figure 12–16b). As the blocks are being laid, ties are installed which link the jamb or frame to the block wall. A spreader must be kept within the doorjamb to maintain proper spacing. It may be necessary to use more than one spreader to maintain uniform spacing of jamb sides. If so, they should be installed as shown in Figure 12–16b. Also note that the reinforcing bars are installed in the hollow next to the doorjamb on both sides and that grout is added to bond both the blocks and the rod.

6. *Window openings.* Sash blocks or open-end blocks should be used when establishing the opening for a window. Sash block is generally used for metal windows and open-end block is used for wood-framed windows. The metal sash is grouted into the slot provided in the block (Figure 12–17a). A nominal 2 × 4 or 2 × 3 can be installed in an open-end block and grouted in position. Frequently, the mason drives 16d common galvanized nails into the back of the 2 × 4 to ensure stability and anchorage in the grout (Figure 12–17b).

7. *Lintels.* Lintels (Figures 12–7 and 12–8) are easily made with the lintel block. Instructions were given on page 196 for the soffit and

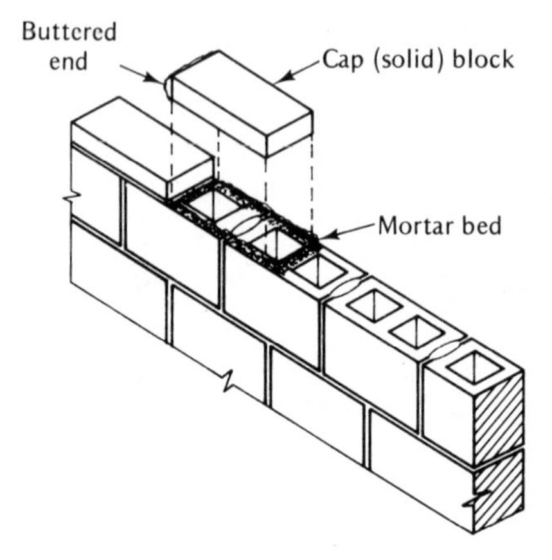

Figure 12–18 Solid Blocks (Caps)

support requirements. After these members are in place, the blocks are mortared into place. It is important while mortaring to clean all excess mortar from within the block. Then a small quantity of grout is placed into the cavity; rods are placed over the grout and separated, and finally grout is filled to the top of the block and struck flush.

8. *Bonding the top course.* The top one or two courses are bonded with reinforcing rods and grout. Figure 12–8 illustrated the placements. A bond beam block is especially designed for this use. It is laid in customary fashion, then filled with grout and two rods encapsulated within the grout.

Since it is not usual to fill all cavities in the wall with grout, a screen mesh (Figure 12–2) must be laid over the course below the bond-beam-block course. This screen prevents the grout from settling to the footing, yet aids in strengthening the wall, since some grout penetrates the screen and forms clinkers.

9. *Cap blocks.* Many plans call for the top course to be solid. A solid, cap block can be used. This block does not provide added strength to the wall; it just covers the holes in the blocks below it. A cap block can be laid over a bond beam block, and when done, it provides a finished appearance to a wall. It is especially useful in finishing a basement wall and in finishing a garden wall (Figure 12–18).

10. *Striking joints.* Various methods for striking joints were illustrated in Chapter 9. The two most commonly used in block laying are the concave and the V joint. Both meet the standards for compressing the mortar to make it watertight. These joints, when pointed, also meet the moisture-penetration and erosion requirements set by industry standards (Figure 12–19).

Figure 12–19 Striking Joints *(Courtesy of Marshalltown Tool Co.)*

BLOCK VARIETIES AND EFFECTS

Regular masonry blocks, decorator blocks, and slump blocks are all made in a variety of sizes. Some of the available patterns and shapes are shown in Figures 12–20 and 12–21.

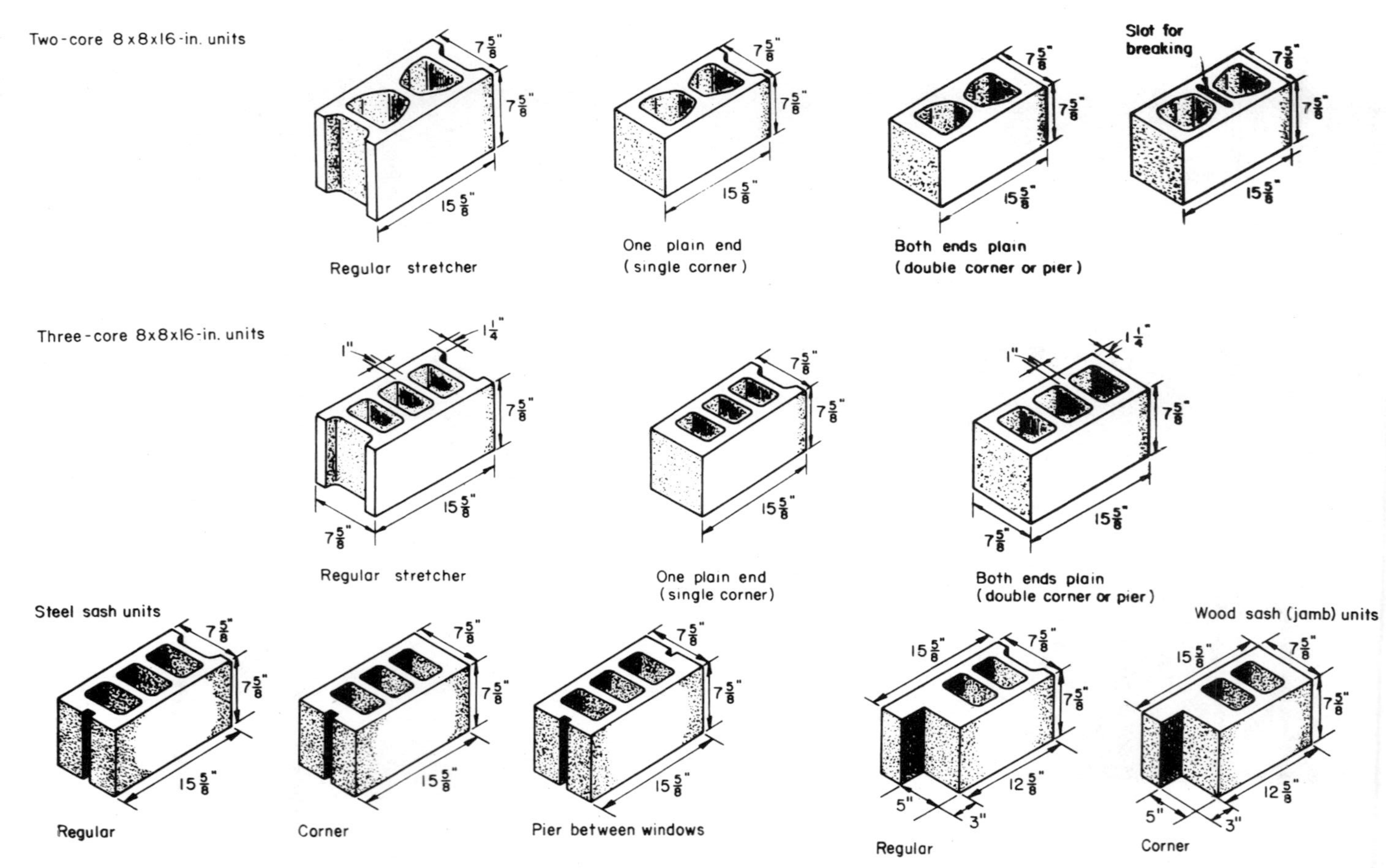
Two-core 8x8x16-in. units
Regular stretcher
One plain end
(single corner)
Both ends plain
(double corner or pier)
Slot for
breaking
Three-core 8x8x16-in. units
Regular stretcher
One plain end
(single corner)
Both ends plain
(double corner or pier)
Steel sash units
Regular
Corner
Pier between windows
Wood sash (jamb) units
Regular
Corner
7 5/8"
15 5/8"
12 5/8"
1"
1 1/4"
5"
3"

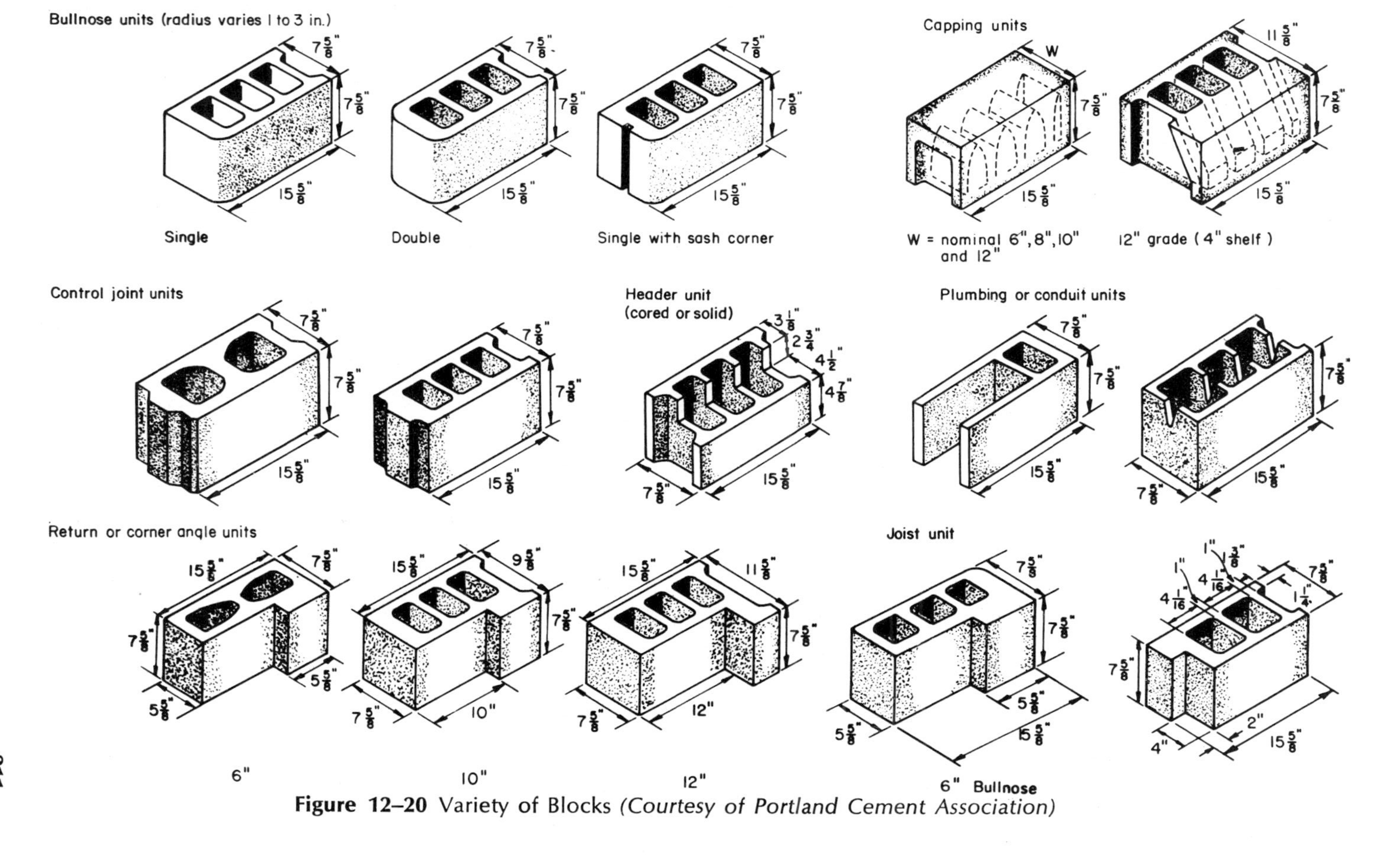

Figure 12–20 Variety of Blocks *(Courtesy of Portland Cement Association)*

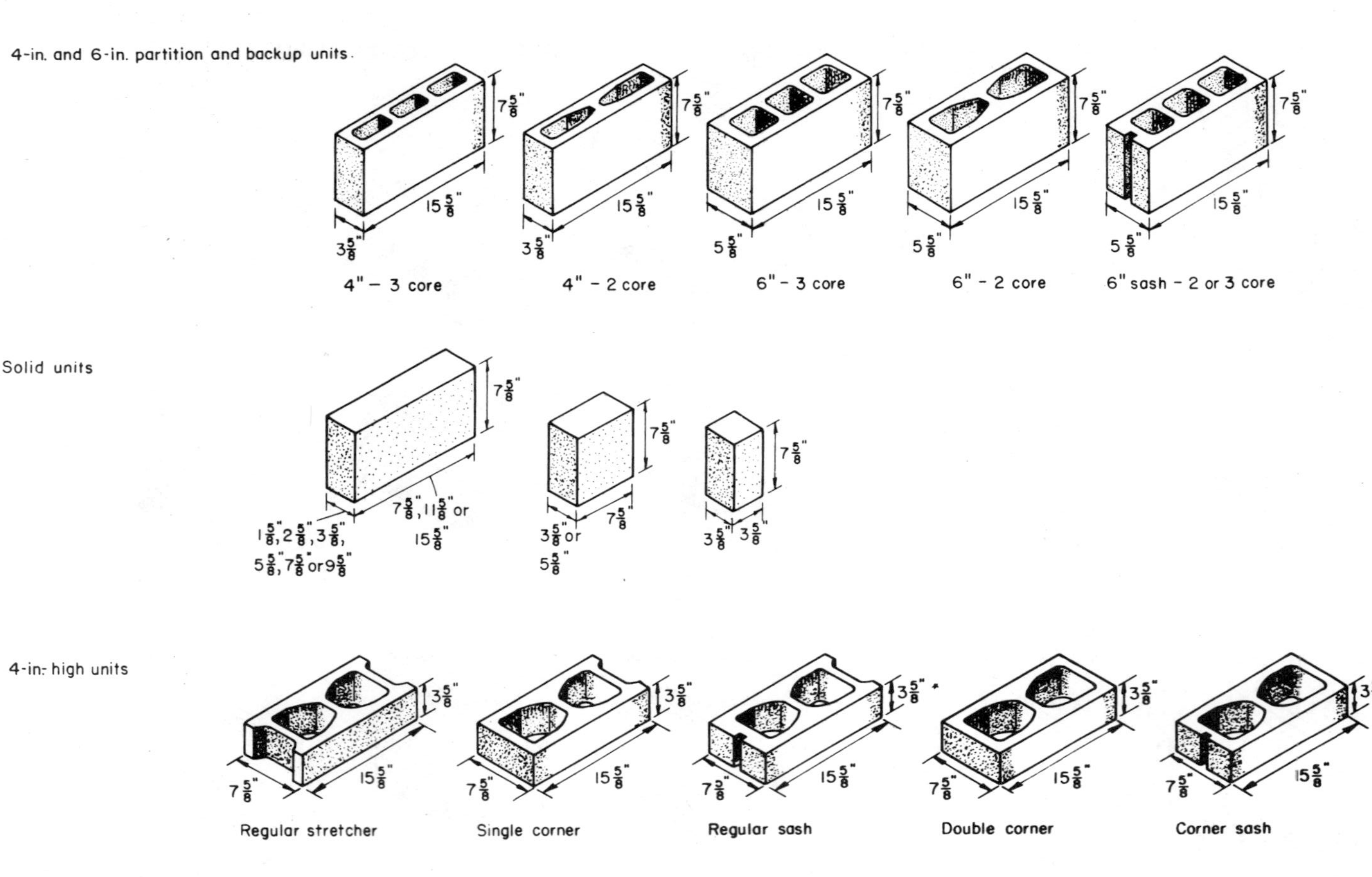
4-in. and 6-in. partition and backup units.
4" – 3 core
4" – 2 core
6" – 3 core
6" – 2 core
6" sash – 2 or 3 core
Solid units
4-in. high units
Regular stretcher
Single corner
Regular sash
Double corner
Corner sash

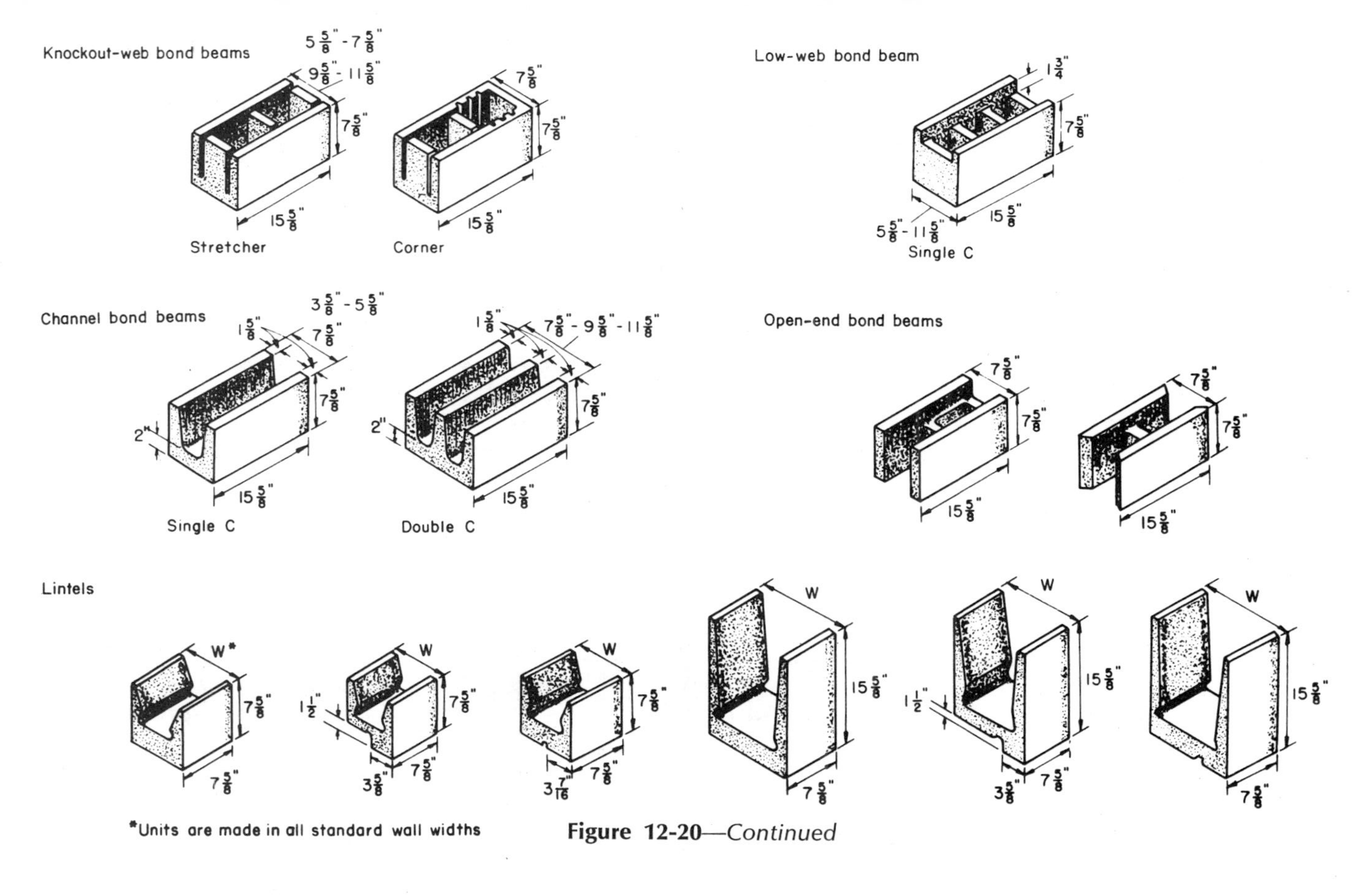

Figure 12-20—*Continued*

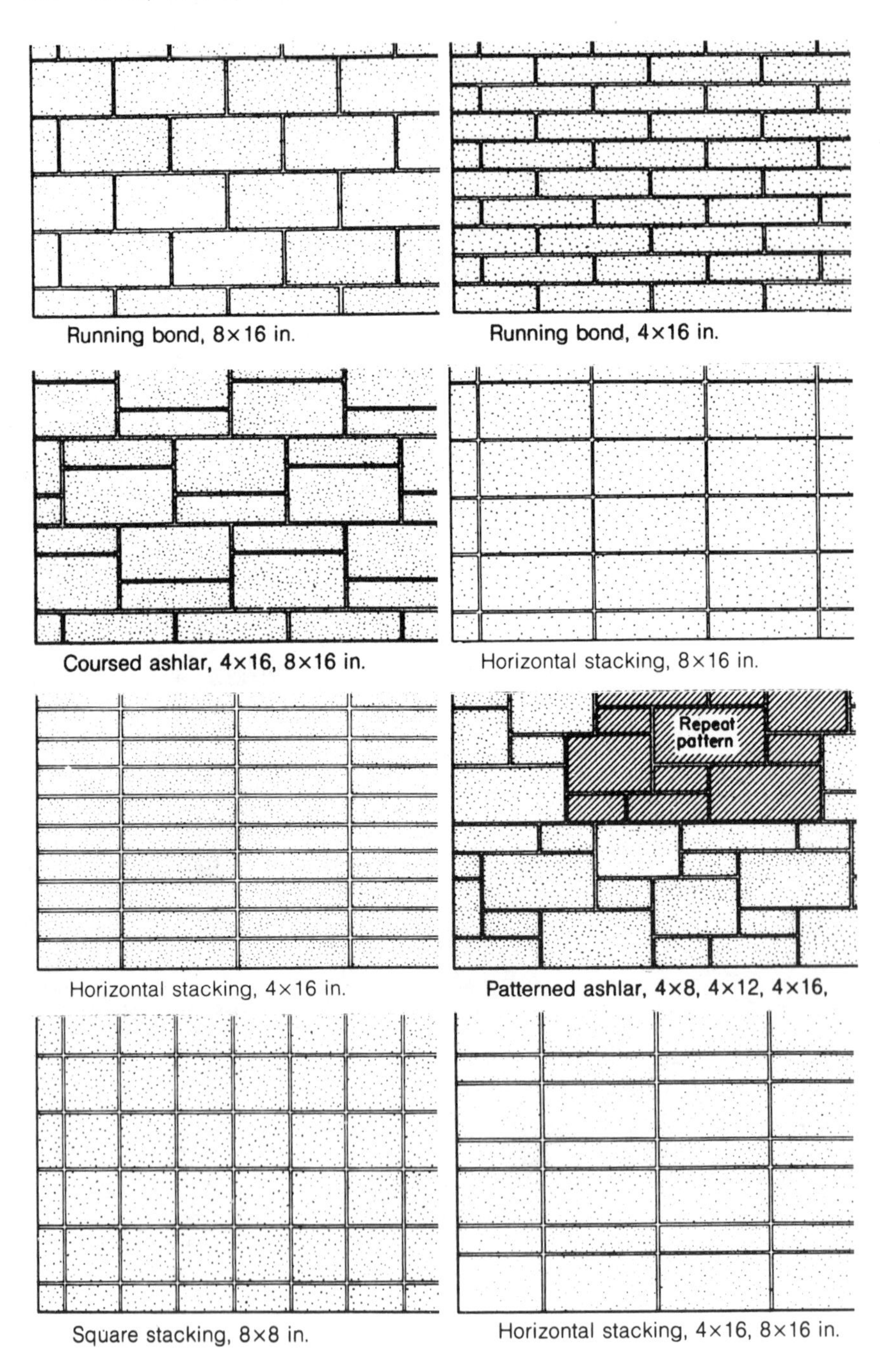

Figure 12–21 Variety of Wall Patterns *(Courtesy of Portland Cement Association)*

INSPECTION OF BLOCK-WALL CONSTRUCTION

Inspection	*Satisfactory/N.A.*	*Unsatisfactory*
Was the proper mortar used?	______	______
Are vertical and horizontal alignments proper, plumb, and level?	______	______
Were all reinforcement rods installed and grout added?	______	______
Were lintels constructed properly and reinforced?	______	______
Were all walls bonded near the top?	______	______
Were all joints pointed or struck uniformly?	______	______
Were all anchor bolts securely embedded in grout?	______	______

QUESTIONS

1. What is a running bond?
2. List four advantages of using blocks to build a wall.
3. Where is a J block commonly used?
4. Which plan in a set of blueprints would show you foundation details and how blocks are to be used?
5. What is the function of wire mesh placed between courses of block?
6. If a J block is used around the perimeter of the foundation, would its highest lip be set at the finished floor height?
7. What is a masonry ledge?
8. What type of block is used in window or door lintel construction? Explain how it is used.
9. Why is it important to use a bond beam near or at the top of the wall?
10. List at least three places where vertical reinforcement must be included in a block wall.
11. Explain how pilasters are made from blocks and grout.

12. Where is the customary starting point for laying each course of block?
13. Why does the mason make a furrow in his mortar before laying the first course of block?
14. Does having the doorjamb installed prior to laying the block aid the mason in any way? How?
15. What function does a cap block perform when installed on a wall? Does it add strength to the wall?

Chapter 13

Rubble-Stone Walls

Anchor wires 16-gauge galvanized lengths of wire twisted around wire mesh and extended over the stone and embedded in mortar.
Ashlar a squared block of building stone. A wall made of squared building stones in the ashlar pattern.
Grout thinned mortar or a thin consistency of 1:3:2 cement/sand/pea gravel mix.
Header bond a header stone installed so as to unite all stones surrounding it.
Header stone a stone laid at right angles to the face stones on a horizontal plane.
Lateral pressure side-to-side pressure (e.g., from wind or earth).
Pointing troweling mortar into a joint after stones (or masonry units) are laid.
Polyvinyl plastic sheet goods sold in rolls and used as insulation.
Rubble rough fragments of broken stone either naturally formed or quarried; used in masonry.
Seismic pertaining to earthquakes.
Tar paper paper impregnated with tar, used for insulation; also called "felt" and "building paper."

OBJECTIVE—INTRODUCTION

Quarry or field rubble stone is frequently used for retaining walls or veneering one or more walls on a house. Usually, it is used in areas where it is readily available. However, sometimes it is trucked to the site at considerable expense to the buyer. If the stone is used in a retaining wall, it may be left in its natural shape and size. But if it is used to veneer a wall, it is generally cut to size, especially thickness. Then, too, if an ashlar arrangement is planned, the stones must be cut to fairly uniform rectangular shapes. If left in their natural shape, a random pattern is established. All these variables, plus several more that will be defined later in the chapter, lead to its objective: *to be able to plan for and understand the requirements necessary for stone-wall construction.*

Different requirements must be employed in the construction of a retaining wall made of stone from those used in veneering; therefore, each subject is examined separately. First, stone used in a retaining wall is examined from the point of view of the function of the wall, foundation of the wall, stone placement, and mortaring vs. not mortaring the stones. Next, stone used as veneer on a wall is examined from the viewpoint of footing needs, size of stone, tying needs, mortar needs, and patterns. The chapter closes with an inspection checklist.

STONE USED AS AN EARTH RETAINING WALL

Function of a Stone Wall

As with all retaining walls, a stone wall has the primary function of supporting earth to prevent erosion or to eliminate a steep slope that would be hard to maintain. It can be built with quarry rubble or with field rubble stone. More often than not it is built with fieldstone that is plentiful in the area. However, near quarries, quarry rubble stone could be used.

The wall must be able to withstand lateral pressure, shifting of the ground, and attempts at erosion by rainfall. The lateral pressure is most severe during the period following backfilling, so extreme caution must be exercised as backfill is added. In seismic areas the wall may be subjected to earthquake-created pressures and forces, which are very likely to create visible damage. This may be the deciding factor in whether or not to use mortar. Finally, erosion by rainfall and ground seepage may erode the wall unless planned for. Again, this factor may influence the use of mortar, or it may be the reason that drainage is included in the wall's construction.

Foundation for the Wall

Figure 13–1 shows the two possibilities for wall foundations. Figure 13–1a shows a typical poured-concrete foundation (footing), which may or may not be reinforced. The size of the footing, its width and thickness, are determined by the type of stone used. Heavier stone, such as granite, requires a more substantial footing than does fieldstone or sandstone.

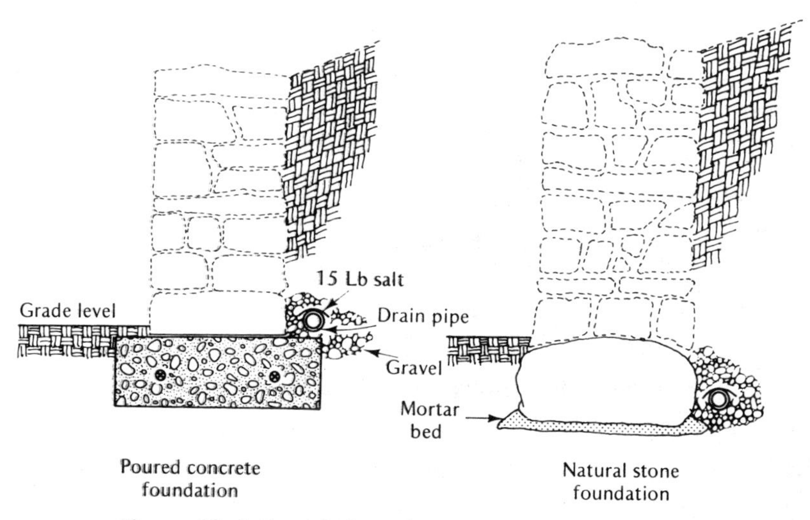

Figure 13–1 Foundations for Stone Retaining Walls

The foundation can also be made from stone if the larger pieces are positioned as shown in Figure 13–1b and embedded in 2 in. of mortar. Smaller stones are usually placed on top of the larger, foundation stones.

Notice that if drainage is expected to be a problem, tile and gravel are placed behind the foundation or first course of stone. Moisture in the earth finds its way to the wall, drains down the wall, and is carried away in the drainage tile. Also notice that both foundations are below the finished grade level. This anchors them well and aids in controlling shifting, should there be any.

Stone Placement

Figure 13–1 also illustrates two significant points in laying the stone. Every three or four courses, *header stones* must be installed. These unite the lower and upper courses, since it is usual to make the wall at least two stones deep. The other point that should be considered is that the *face* or *outer surface* of the wall should slope toward the bank of earth at the top from a few degrees to 10 or more, depending upon the effect desired.

Although the drawing does not show it, the inside stones may be very irregular and protrude quite a distance into the bank. This is,

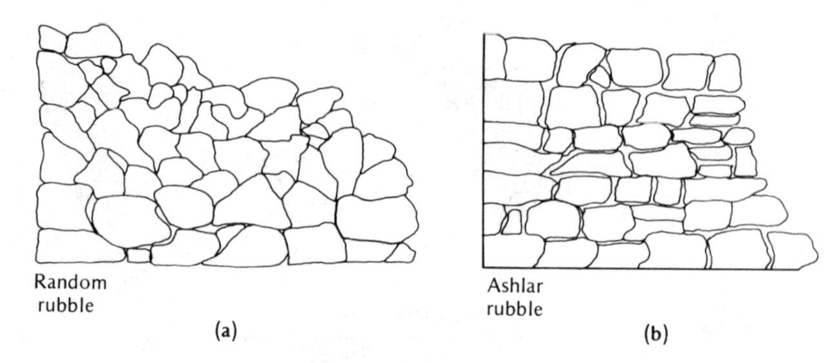

Figure 13–2 Laying Stone in a Retaining Wall

for the most part, desirable, since each stone acts as an anchor for the wall.

Figure 13–2 shows the two arrangements of stone laying. Figure 13–2a illustrates a random rubble-stone pattern. In this method, stones are laid in any pattern that seems to fit well. Obviously, header stones are randomly placed, but they should be placed approximating one each 2 to 4 ft^2. Chips and small stones are used to fill cracks and crevices between larger stone. In contrast, Figure 13–2b shows rubble stone laid in courses. To do this means that the stones are shaped with a brick hammer and chisel to approximate rectangles before they are laid.

Keeping the wall face perpendicular or sloped away may be accomplished in one of two ways. If accuracy is not demanded, sighting with the human eye or by use of a plumb bob is sufficient. The wall face will be rather uneven and very natural.

If, on the other hand, a more formal effect is needed, a mason's line, level, and plumb bob are used to maintain stone face alignment.

Randomly shaped or very rectangular quarry stone should be carefully selected and laid out so that the face of the wall can be maintained in line during construction. Even with this restriction, either pattern previously described can be used in a more formally designed wall.

Mortar vs. No Mortar

Mortar secures each stone to its neighbor; therefore, it must be the bed for the stone and also must complete the bonding of stones. Masonry mortar and sand should be used with tap water to make a mixture. The lime in the masonry mortar does not stain stone, but if staining could be a problem, white portland cement should be used instead of regular portland cement.

A mortar bed should be laid thick enough to allow the entire stone to be embedded in it. If the joint between stones exceeds 1½ in., small chips or pieces of stone should be pushed into the mortar to fill the void between stones. Mortar should be used on all vertical joints and between front and rear stones. The mortar also aids in maintaining wall alignment.

Nonmortared retaining walls are frequently built. The foundation stones are laid in mortar beds. From that point on, each stone is laid dry upon each other. As each layer or course is laid, earth is backfilled behind the courses and tamped. This aids in securing the wall.

Since there is no mortar, stones must support and anchor stones, so small stones aid in supporting larger ones. The header bonds become very important, because they unite front and back rows of stone from above as well as below.

Natural drainage is evident in this type of wall construction. Water seeps out of the wall between the stones, especially if some of the stones tilt forward slightly. If all tilted backward, water would be trapped behind the wall and drainage pipe and gravel would need to be installed to carry it off. If no seepage or minimal seepage is desired, drainage pipe and gravel should be installed.

STONE VENEER ON A HOUSE WALL

The simplicity and general ease of building a stone retaining wall are replaced by a more scientific approach when stone is used to veneer a house. First, its thickness is limited to from 2 to 11 inches. When the desired thickness is selected, all stones must be cut to this approximate thickness. Next, because a single tier or wythe is used which of itself contains little stability, it must be securely anchored to the wall. Also, the foundation must be adequate to support the weight of the stone (the thicker the stone, the greater the weight, of course). Finally, grout is added to unify stone and wire mesh or appropriate backing.

Footing Needs

The footing must be wider than normal to accommodate the wall, sheathing, wire mesh, a minimum of 1 in. for grout, the thickness of the stone, and 4 in. (10 cm) for overhang, as Figure 13–3a shows. The

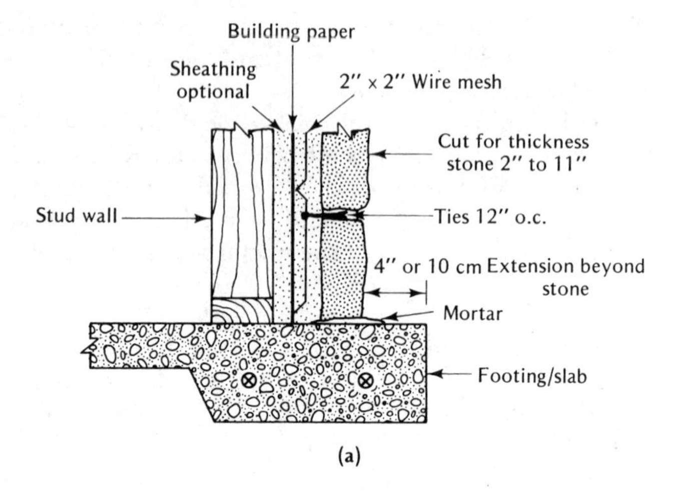

(a)

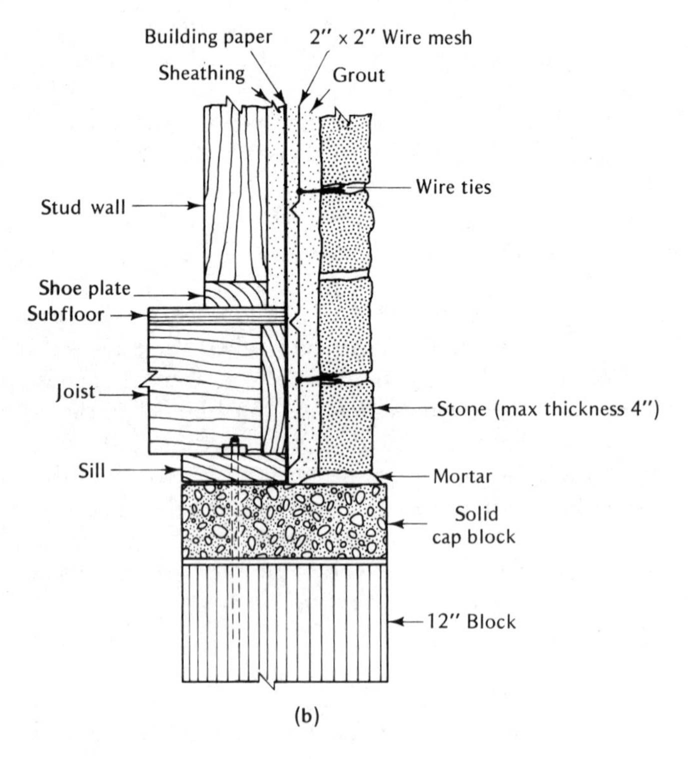

(b)

Figure 13–3 Stone Veneer Section Drawings

weight, pressure, and force of the stone is calculated by the architect so that an adequate footing thickness is planned. The data given in elevation section or detail plans similar to Figure 13–3a (but more complete) must be faithfully complied with.

If stone is used on a structure that has a basement or crawl space, narrow stones up to 4 in. in thickness are usually used. The reason for the limitation on thickness is shown in Figure 13–3b, where 12-in. blocks are used for the foundation, allowing a 4-in. ledge for stone to be set upon.

Planning Stone Placement

The selection of either random or coursed (ashlar) pattern is used in veneering. The larger stones are generally installed near the base, and smaller ones from midheight up. That is not to say that larger stones should not be used on the upper half. Rather, their frequency should be minimized and they should be scattered throughout the wall. In like manner, medium-sized and smaller stones should be interwoven with the larger stones at the base to help offset any severe concentration of mass. The best patterned walls have a balanced appearance.

It may be desirable to first lay a pattern of stones on the ground. For example, a square or rectangular area the size of the wall's length and height can be measured off on the ground. One side is designated as the base. From this point on, stones are selected, cut, and placed along the base line and from perimeter to perimeter until the area is filled. Then an overview from several yards back can reveal the pattern and effect. If pleasing and proper, stones can be installed on the wall one at a time.

Anchoring Arrangements

Assume that the following is complete: sheathing, tar paper, or polyvinyl are installed on the studs and securely nailed. A 2 × 2 wire mesh is cut and nailed to the wall with galvanized nails in such a manner that there is spacing between tar paper and mesh except at nailing points. (*Note:* Some local codes permit exclusion of plywood or other sheathing.) With the wire mesh nailed securely at 4 in. o.c. in every stud, the stone can be laid. After laying the first course, a set of wire ties needs to be inserted and twisted around the wire mesh and allowed to extend onto the stone. These wires should form a matrix of 12 in. × 12 in. (30 cm), as Figure 13–4 shows. Figure 13–4 shows several layers of stone mortared in place with the wire ties added.

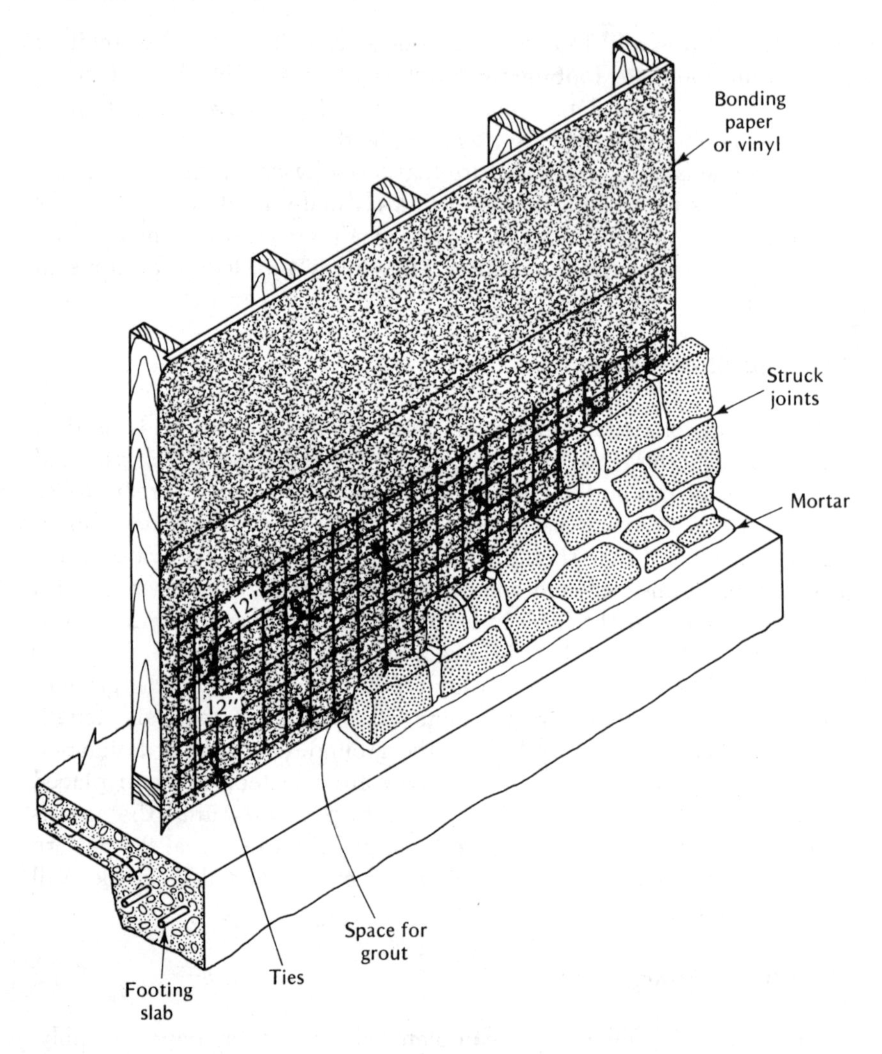

Figure 13–4 Veneering Stone and Anchoring

The final phase of anchoring requires the addition of grout between the stones' inner or back surfaces and the wall. The grout encapsulates the wire mesh and bonds to the stone, thus unifying the structure. The grout may be added either as the wall is being laid, say each 2 ft, or it may be added at one time after the wall is raised. If added periodically, it should be kept several inches below the last course of stone. If done last, it should be placed carefully to avoid displacement of stone.

Mortar

The mortar described earlier should be used for this veneering operation, also, but where splashing of mortar may not have been a critical factor in building a retaining wall, it is in veneering. Therefore, care must be used when placing mortar for either *bed* or *vertical* joints. Proper lifting motion with the trowel aids in controlling splashing, as does a clean trowel. Brushing with water washes away stains of mortar.

It may be necessary to use the pointing technique in joints, especially if cleaning has disturbed them. Additional mortar is added as the pointing is done. In this manner compression joints are made with the pointing tool, and a sound water-resistant finish is added to all joints.

INSPECTION OF STONE MASONRY

Inspection	*Satisfactory/N.A.*	*Unsatisfactory*
Has a proper foundation been laid?	________	________
Have stones of reasonably uniform size been selected or cut?	________	________
Has the proper mortar type and composition been used?	________	________
In a retaining wall:		
a. Were stones laid with every 2 ft of vertical wall height?	________	________
b. Was the wall tilted toward the earth several degrees?	________	________
c. Was drainage included?	________	________
In a veneered wall:		
a. Was the wire mesh installed properly?	________	________

Inspection	*Satisfactory/N.A.*	*Unsatisfactory*
b. Were stones laid in proper mortar beds?	______	______
c. Was grout used to bond stone to wire mesh?	______	______
d. Were joints struck properly?	______	______
e. Were stone faces cleaned of excess mortar and stains?	______	______

QUESTIONS

1. What does "ashlar" mean?
2. What is a header stone?
3. Define rubble.
4. During what point in the construction of a rubble-stone retaining wall will the wall experience the greatest lateral pressure?
5. Name the two methods of preparing a footing/foundation for a rubble-stone retaining wall.
6. Under what conditions should gravel and drainage tile be used behind a rubble-stone wall?
7. How frequently must header stones be installed in a wall?
8. If a wall is raised without mortar, what features should it have that allow for water seepage?
9. What modifications to footings need to be made if stone is used on a house as veneer?
10. Why is it desirable to lay the stones in a pattern on the ground prior to installing them on a wall?
11. Explain the method used to anchor stones to a house wall.
12. What function does grout have in building a veneered wall of stone?
13. How are compression joints made between stones?

Section III

Artistic Forms of Masonry

Section III has only two chapters. Contrasted with the first two sections, that may seem very short, but, of course, the same materials that were examined before are also used in this section, but here the emphasis is on creating artistic forms.

Naturally, there is a need for readers to know how to go about developing their own ideas, then to create the finished product, especially when there is no architect available to draw up a blueprint and specifications. To aid readers in developing ideas, Chapter 14 presents a seven-step flowchart. As its name implies, a flowchart shows the user where to start and what steps to follow during project development. So that each step will be understood, we have provided explanations and included examples.

Chapter 14 may be sufficient for some readers and they can take off on their own. But for other readers, this will be difficult, and Chapter 15 is for them. There several projects are developed in detail so that the reader can relate the ideas and sequences used to create an object. For example, projects 15–1 and 15–2 use wood forms to help create a flower box and column or post. In contrast, project 15–3 uses a free-style form made of sand. Through study of these varying techniques, readers will learn how to go about selecting the methods best suited to their needs.

For the most part, the mixture of cement, sand, and water used for free-style-formed projects must be made very dry. Sufficient water must be added to allow the cementing process to start and to continue. But the mixture should be so dry that the granules of coated sand appear to fall apart. If this technique is followed, chances of success during forming are greatly improved. To ensure that proper curing takes place for 7 days, the project must be watered periodically and, if possible, covered with plastic sheet goods.

Where subassemblies are to be mortared together and in place, standard mortar mixes should be used. All joints should be compressed, even those that are to remain flush.

Chapter 14

Masonry Sculpture Fundamentals

Artwork a sculptured or created object.
Color admixtures of natural coloring agents included in concrete or mortar.
Free forms any form (made of wire or sand, for example) that details shape and design.
Natural materials stone, wood, remains of other structures.
Plastic mixture concrete or mortar prior to setting.
Texturizing creating a particular finish, such as brushed, smoothed, etched, pockmarked.
Time-line plan method of organizing time in sequential progression.
Work means artwork or sculptured form.

OBJECTIVES—INTRODUCTION

Masonry sculpture fundamentals are significantly different from the standard fundamentals used in the construction of masonry projects. The floor and elevation plans customarily needed for general masonry work are not appropriated for sculpture work. However for a project to be completed successfully, it is equally important to have a general plan as well as a set of plans or working drawings. This chapter provides suggestions and requirements for all sculptured art forms, and this leads to the first objective: *to be able to formulate a plan for art forms when using concrete and related materials.* Several of the key points to examine are ideas, possibilities of making the artwork, sequence of planning to be used, and estimating the supplies that will be needed.

It is not sufficient to be able to develop the plan; it is also important that the plan be used. To this end, there is the second objective of this chapter: *to be able to translate an art plan into sound working methods.* Within this idea several aspects of construction are examined including building the forms, mixing and applying the mortar or concrete, and creating textures or finishes.

ELEMENTS OF A PLAN FOR MASONRY ART

Ideas

The idea for a project may arise either from your own basic concept or by adapting someone else's ideas. First, one can use one's imagination. If the thought is carried far enough, many details, including size, shape, peculiar angles or curves, color, texture, and combinations of materials can be determined.

A second source of ideas is another person. A casual remark may trigger an idea. Or a more detailed description may provide the nucleus of a plan.

A third source is experimentation. Models can be made and expanded upon; or various approaches can be tried until a satisfactory set of conditions is found.

During the exploratory phase of developing basic concepts, organization is important. The organization indicated here consists of listing the principal characteristics of the idea, followed by outlining concepts in work to be done or planned for. If the principal characteristics are listed, there need not be a fixed order; any order may be used since the aim is to get all ideas down in note form. When sufficient notes are made, a preliminary ordering can be undertaken. This may be as simple as order-of-importance sorting, or a more complex arrangement may be made.

If a basic concept is difficult to achieve, the alternative of using someone else's idea should be chosen. First, check the literature for a *look-a-like item.* The best resources are magazines, books, and manufacturers' literature—or you can use the projects detailed in Chapter 15.

If the look-a-like item is to be used, it must be examined critically: the results of dimensional changes must be noted, as well as changes in materials that could affect the texture, quality, and durability of the work or its esthetic value. Finally, how does the adapted art form fit into the intended location? Will it overpower the surroundings or be lost in them?

It is not necessary that the art form serve a purpose, although it can. Therefore, purpose should not be the primary factor, but it could very well be very important.

Making the Art Form

There are two broad categories that must be considered before undertaking any work of this type, and it is well to examine them during the planning phase. First is your ability to carry through the project. The materials selected may be quite heavy. Recall that 1 ton of materials is needed to make 1 cubic yard of concrete. Also, note that cement blocks weight up to 60 lb each. The location may also present physical limitations. It is also conceivable that a person could have real difficulty formulating a plan because of a lack of understanding of the properties of the materials to be used. It takes an effort to organize a plan that considers every practical task or anticipated problem.

The second consideration involves *resources availability.* If the work includes natural materials: are they locally available, natural to the area, and economical in cost? If not, making the work becomes more difficult. If the work includes manmade products, the same question can be asked. If either of these is an obstacle, examine the possibilities of using alternative materials that produce the same effect. It is also possible that if local materials are incorporated, the total effect will be a work of art that fits the location ideally.

When a decision has been made that there is a genuine possibility of creating the artwork, the practical parts of the plan should be developed.

Sequence of Planning To Be Used

A flowchart approach is a clear method of showing a sequence of operations. Flowcharts effectively display sequences that are linear, those that must be followed in order, and those that allow branching out. In the effort of planning a project of this sort, this chapter describes the stages of operations in a flowchart (Figure 14–1).

The first three stages have already been defined. The last four stages are examined further later in the chapter. Even though the flowchart shows how to organize the project and carry it through to a successful conclusion, a fundamental part of the plan has not been included. Supplies and time are needed.

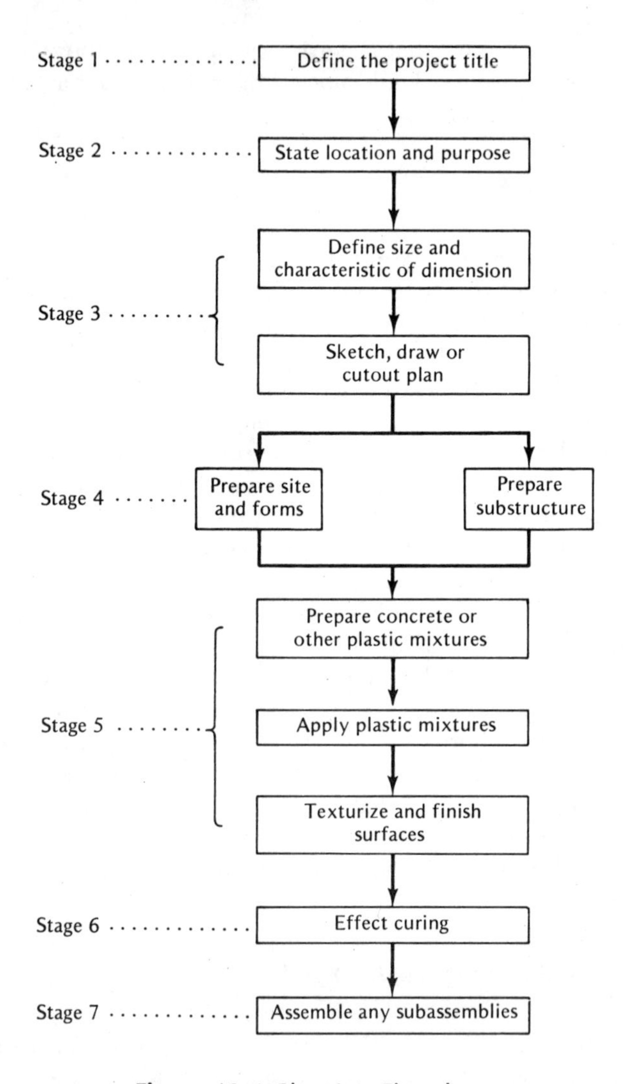
Stage 1
Define the project title
Stage 2
State location and purpose
Define size and characteristic of dimension
Stage 3
Sketch, draw or cutout plan
Stage 4
Prepare site and forms
Prepare substructure
Prepare concrete or other plastic mixtures
Stage 5
Apply plastic mixtures
Texturize and finish surfaces
Stage 6
Effect curing
Stage 7
Assemble any subassemblies

Figure 14–1 Planning Flowchart

Estimating Supplies and Time Needed

There are four basic steps needed for each estimating phase in any project.

1. Make a list of all materials by size, weight, volume, and quantity. Include sand, blocks, brick, wire mesh, reinforcement rods, form materials, and the like.

2. Group all like materials to obtain the totals.
3. Make a list of all tools needed to make the work easy and professional. Good-quality tools (Appendix A) are a wise investment. Once bought, take very good care of them.
4. Call a local lumberyard, building supply house, or masonry supply company for estimates. They will gladly figure quantities and price their costs. They may also deliver free of charge.

Time is difficult to estimate in any planned undertaking. If the project is to be built in more than one session, planning becomes especially important. Some decision making must take place so that sufficient time is alloted to each phase of construction. For example, if subassemblies are needed and forms must be made for each, time could be allocated as follows:

1. Build forms; estimate 3 hours (a Saturday morning).
2. Pour forms with a mortar or concrete mixture; estimate 2 hours (the second Saturday).
3. Allow 7 days for curing.
4. Mortar subassemblies together; estimate 3 hours (the third Saturday).
5. Finishing, texturizing, and cleanup; estimate 4 hours (Sunday or the fourth Saturday).

Of course, a wider variety of tasks might be involved and, if so, the time-line plan would be very detailed. Nevertheless, the idea is to plan the work to be accomplished in each phase. Also of serious concern in time allocation is the weather. Recall from Chapter 1 that extremes of weather must be avoided. Careful check of the local papers for weather predictions can aid in the planning effort.

TRANSLATING THE PLAN TO A PHYSICAL OBJECT

Stages 4 through 7 in the flowchart identify the elements of the plan that involve the actual construction and development of the artwork. These four stages define the general tasks needed to satisfy the second objective of this chapter. Since the brief terms in each block of the flowchart may not be completely clear, explanations follow that should more clearly define each one.

Defining the Form(s) Needed for Support (Stage 4)

An in-depth understanding must be established from available data, knowledge, and insight as to the type and complexity of the form needed to support the work. This form may be either an integral part of the artwork itself, such as reinforcing wire, or it may be a form that develops shape and dimension, such as a footing form. Forms made as an integral part of the artwork itself may be fabricated from chicken wire, 2 × 4s, or 4 × 4s, mesh, steel rods, or a combination of these items where concrete is used as the covering medium. These materials may also be incorporated to join unlike materials where mortar is the bonding agent. Most applications of the materials described here fall into the broad category of *free forms*.

One type of form frequently used is made of sand. Sand, usually masonry, is pressed into a shape representing the shape of the artwork. The shape may be strict, as in rectangles or squares, or more natural, as in curves and irregular patterns. In either case the sand must be wet enough to stay in place. Once shaped, it is desirable to spray it with lacquer or varnish so that it retains its shape and the covering of the lacquer makes a slightly smoother texture. From this point on, the material (such as concrete or mortar) is applied very carefully.

Besides these two methods, there is the wooden form. It may be constructed of any suitably smoothed wood. If curvature is needed, several layers of very thin pieces are usually used because of their ease in bending. Whether using thin layers or thick boards, the form must be well braced where the pressure of plastic materials is thought to be concentrated.

Finally, a combination of all of these methods may be employed at one time. For example, sand may form a basic internal hollow, a wood form develops the outer shape, and wire mesh provides the internal support for the plastic mixture.

Preparing and Using the Plastic Mixture (Stages 5 and 6)

Some mention has been made of concrete and mortar but not enough for complete understanding. Frequently, the plastic mixture of mortar or concrete must be made differently from the way it is normally made. A dry mixture of portland cement and sand is sometimes used. This mixture represents a sand whose moisture equates to saturation and where little if any additional water is used in the mixture. Here experimentation plays an important role. Suppose that a concrete flowerpot is to be made. There must be an inner and an outer form. The normal consistency of mortar would allow for easy filling of the form but would be difficult to control. On the other hand, a dry mixture could be placed into the form in small quantities and compacted thoroughly. Curing could easily be controlled if the project were enclosed in plastic and allowed to remain for 7 days. Before placing the plastic over the project, the outer form could be removed and in some cases the inner form as well.

It is possible to texturize the outer or inner surfaces of the project just after molding and while still uncured, but great care must be used. Texturizing may also be accomplished after the curing stage.

Assembly (Stage 7)

The final stage, 7, implies that more than one subassembly has been constructed or purchased. It also implies that when sufficient subassemblies are ready, they should be united. If all subassemblies or units are masonry, uniting them could simply involve making a mortar and using it to form bonding joints. However, if nonmasonry materials are used with masonry, other forms of unions are needed. These might include bolts, straps, or special glues, or expoxies. In the event that this is planned, the type of union must be structurally sound.

QUESTIONS

1. What does texturizing mean in terms of masonry art?
2. Name two different sources of ideas for masonry art.
3. Why is it important that a preliminary ordering of collected notes and data be made?
4. Name four possible sources of look-a-like items.
5. If a look-a-like item is to be modified, what are some of the key factors that must be considered?
6. What are two major obstacles to an artwork project?
7. How do you develop an estimate of materials?
8. What are some of the materials that are used as free form?
9. What is a dry mixture of concrete?
10. What is a usual method of joining subassemblies made of concrete materials?

Chapter 15

Projects in Masonry - Art Forms

The projects included in this chapter are accurately detailed so that they may be built as specified. Two of the projects use wooden forms to create the desired shape. One makes use of the sand-form technique, and two make use of masonry units and other materials.

Each project is developed with the same format so that builders can recognize how to organize their own designs. The project elements follow the flowcharting shown in Chapter 14.

The *description* should develop a word picture, including any information that aids in understanding. For example, the parameters of the project can be listed, as well as where and why the project is being built. The types of materials can also be listed and explained. Any special or complicated procedures or techniques can be included. The main reason for developing a description is to put a series of ideas together in one place that will serve as a reminder of what the original objective was.

The *bill of materials* must be developed as accurately as possible, since this list represents the expense of a project. Although it is desirable to complete a project with absolutely no waste, some should be expected. This is especially true where sand, gravel, mortar mix, cement, and perlite or vermiculite are used. Always buy more than expected usage calls for.

The *sequence of creation* consists of several subsections: preparation of plan, preparation of basic units or subassemblies, and finishing tasks. The plan should be used to list all or as many of the parts of the project as come to mind. To be sure that the reader/builder understands what the plan may include, ideas are stated in each project. Obviously, each designer would try to replace these statements with more specific values.

The preparation of basic units or subassemblies must list in a precise sequence the step-by-step procedures that will be used to develop the project. It may take several attempts before the list can be considered to be correct. Experienced builders or masons create these lists mentally; the novice or apprentice must write them down.

The final part of the sequence of creation includes all parts of the project needed to finish the project. As with the previous subsection, this final section must be made in sequential order. It must carefully list the elements that are to be used. Each statement must be as precise and as detailed as necessary.

This explanation of how each project is organized has been provided so that reader/builders can interpret and adapt the concepts for their own use. Now on to some enjoyable projects in masonry.

PROJECT 15–1 FLOWER BOXES

Description

The flower box shown in this project is one made of lightweight concrete that contains perlite as the coarse aggregate (vermiculite could also have been used). It has no reinforcement, since little if any force is expected from the soil it will contain. What makes this flower box simple to make are its forms. The inside of the box is formed with an empty ½-gal plastic ice cream container. The outside form is made of stock 1-in. lumber. The dry mixture of concrete is placed between the two forms. Since the bottom of the box is facing up, the drain hole can be made with a dowel stick, and a depression can be made in the bottom to improve the stability of the box.

Bill of Materials

1	½-gal ice cream container, plastic
1 piece	1 in. × 10 in. × 4 ft lumber
1 piece	quarter-round molding, 3 ft long, cut into 4 7-in. pieces
12	6d common nails
8	3d finish nails
6 qt	sand and perlite
2 qt	white portland cement
1 qt	water

Sequence of Creation (Figure 15–1)

Preparation of the plan:

1. Decide on the size of the flower box.
2. Decide if the outside of the flower box is to be beveled or straight.
3. Decide on the color: add to mixture, apply later, or leave white.
4. The outer form must be constructed of wood; a plastic container can be used for the inner form.
5. The form can be filled by hand while holding the inner and outer pieces in place.

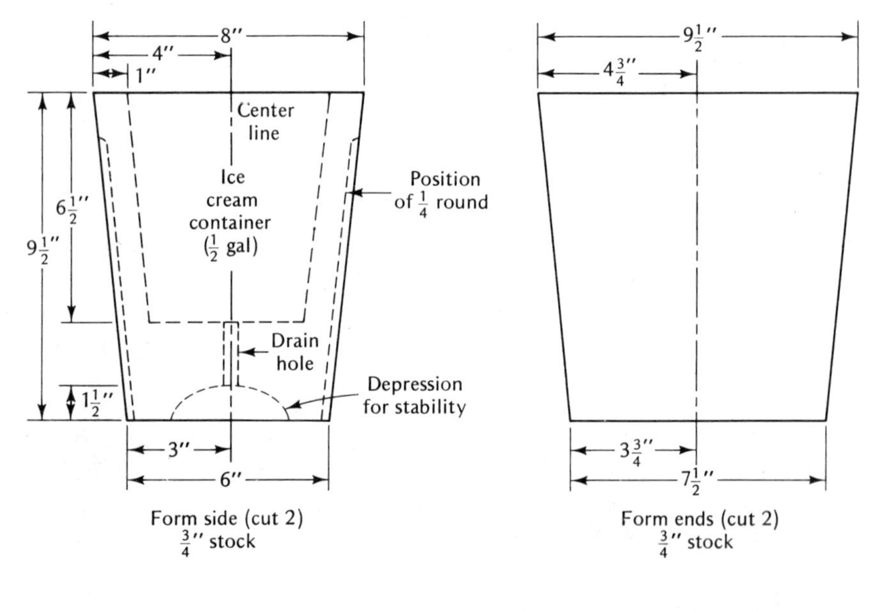

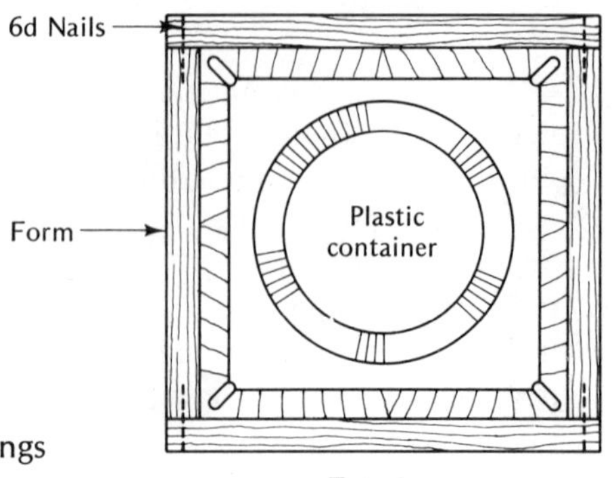

Figure 15–1a Dimensional Drawings for Forms for Flower Boxes

6. A drain hole must be made.
7. A depression made in the bottom will make the box more stable.

Preparation of the base support materials:

1. Measure the diameter of the ice cream container (Figure 15–1a).
2. Add 2 in. to the diameter to establish the total width and length of the upper and lower dimensions of the box.
3. Measure the height of the ice cream container and add 2 to 3 in. for the bottom of the flower box.
4. Lay out the pieces of the four sides that are needed for the outer form. (*Note:* these are beveled; you may want perpendicular sides.)

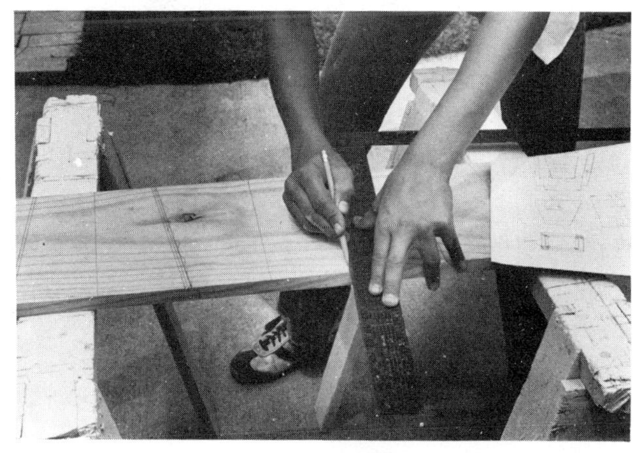

Figure 15–1b Materials Cut, and Being Nailed

5. Cut the pieces and nail them together with 6d common nails (Figure 15–1b).
6. Cut and install quarter-round molding (optional).
7. Make sure that the form is square and, if needed, install iron corner braces.
8. Oil the inside of the form with motor oil.

Placing the mixture and finishing:

1. Mix a dry mixture of cement, perlite, sand, and very little water. (*Note:* Sand should already be saturated.)
2. Place small amounts of the concrete mixture around the inner form and, while holding it centered within the outer form, compact the concrete with a 1 × 2 stick (Figure 15–1c).

Figure 15–1c Filling the Oiled Form

3. When the form is full of concrete and tamped thoroughly, use a trowel and make a depression in the concrete by removing a small quantity. Round this depression with a jar.
4. Next make the drain hole in the concrete with a piece of ½-in. dowel.
5. *Optional:* Lift the outer form and make a second, third, and fourth box if desired (Figure 15–1d).
6. Cover the molded box with a plastic sheet to allow slow curing.
7. Smooth the exterior roughness with a rubbing stone if necessary, plaster with a slurry mixture of cement and sand, and color or leave plain.
8. Extract the ice cream container and plant a flower (Figure 15–1e).

Figure 15–1d Removing Outer Form While Mixture Is Wet

Figure 15–1e Completed Flower Box

PROJECT 15–2 POST AND BASE

Description

The post shown is part of a unit that is used as a mailbox support. It could just as well be used as a post for a lamp, flower stand, birdbath, or fence.

The post itself is made by first making a wooden form, then filling it with a dry mixture of concrete. The post in this project is 38 in. high and includes flutes on two of its four sides. It has a reinforcement bar throughout its length to provide tensile strength. A base made of mortar and bricks is the foundation for this post.

Bill of Materials

2 pieces	3½ in. × 38 in. × ¾ in. form sides
2 pieces	5 in. × 38 in. × ¾ in. form sides
2 pieces	1½ in. × 30 in. quarter-round stock
1 piece	6 in. × 6 in. × ¾ in. form base
1	⅜ in. × 36 in. steel rod
20	6d common nails

Approximately:	2 medium shovelfuls of sand
(concrete mixture)	3 medium shovelfuls of perlite
	2 medium shovelfuls of cement
	2 to 3 quarts of water
10 solid bricks	
approximately:	2 shovelfuls of mortar mix
(for mortar)	6 shovelfuls of sand

Sequence of Creation (Figure 15–2)

Preparation of the plan:

1. Determine the post size to be 3½ in. in diameter by 38 in. long.
2. Decide on the post's design; include fluting on opposite sides.
3. Plan to include one reinforcement rod.
4. Allow some method of anchoring the mailbox.
5. Plan to install the post in a brick base.

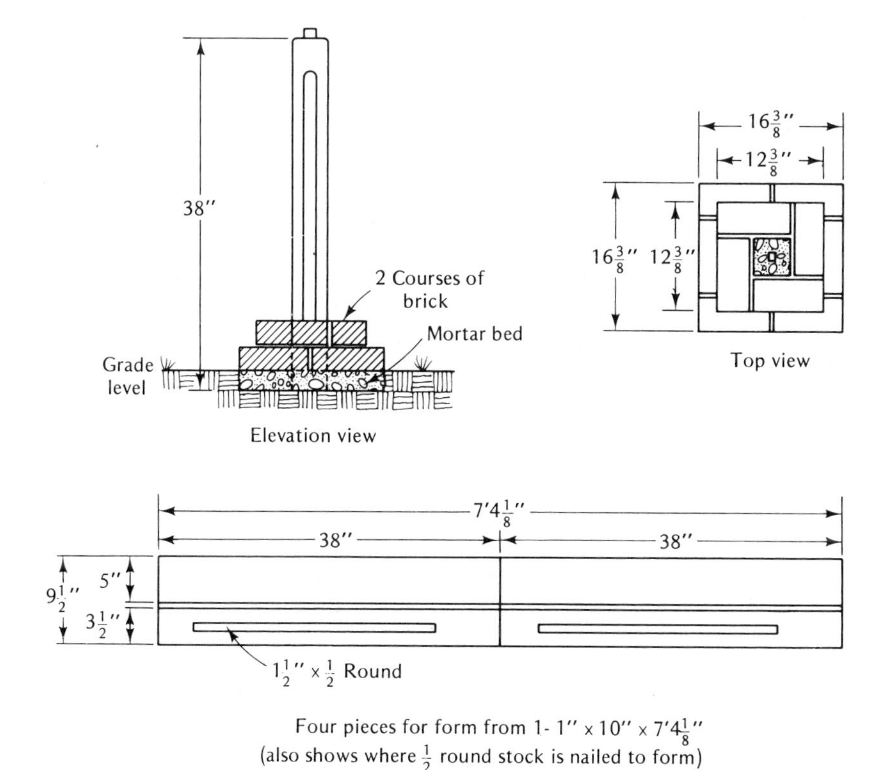

Figure 15–2a Dimensional Drawing for Forms and Bricklaying

Preparation of the post forms and concrete:

1. Lay out the stock for the four sides of the form:
   a. Cut pieces 38 in. long (Figure 15–2a).
   b. Rip two pieces 3½ in. wide and two pieces 5 in. wide.
   c. Cut one additional 6 in. × 6 in. piece.
   d. Cut two 24-in. pieces of half-round and round off the ends.
2. Nail half-round pieces to the centers of the 3½ in. × 38 in. pieces (Figure 15–2b).
3. Oil all four pieces (Figure 15–2c).
4. Nail the four pieces together to form a tube like a square box and then nail the base plate to the form.
5. Mix sand, perlite, and cement in a wheelbarrow, add water, and mix thoroughly.

Figure 15–2b Nailing Half-Round Stock

Figure 15–2c Oiling Forms Before Assembly

Placing concrete and compacting it:

1. Place a small quantity of concrete in the form and compact it with a long 1 × 2 piece of stock until form is one-fourth filled (Figure 15–2d).
2. Install the steel rod.
3. Repeat step 1 until the form is almost full, then add the anchor, which will remain partially exposed.

Figure 15–2d Filling Form and Compacting Concrete

4. Finish filling the form and trowel the top of the concrete flush with the form.
5. Bevel the upper edges of the concrete with the edge of the trowel.
6. Allow the mixture to set for 24 hr while in the form.

Curing the post:

1. Remove the post from the form by taking the form apart.
2. Keep the post wet for 7 days by periodic sprinkling or soaking.

Post base and final assembly:

1. Clear the ground several inches deep where the assembly is to be installed.
2. Lay approximately 2 in. of mortar into the cleared area and set the lower course of brick and post in place (Figure 15–2e). (You may need to brace the post to keep it plumb.) Fill all the joints with mortar.
3. Lay the second course of brick around the post (Figure 15–2f).
4. Strike all the joints and let the mortar set.
5. Install the mailbox on top of the post (Figure 15–2g).

Figure 15–2e Filling Base with Mortar

Figure 15–2f Laying Brick with Post in Place

Figure 15–2g Finished Job

PROJECT 15–3 CONCRETE MUSHROOMS FOR THE GARDEN

Description

The first two projects used wood to form their basic shape; this project uses sand. A small pile of sand is wetted thoroughly, so that (1) it will not draw water from the concrete mixture and (2) so that the sand stays in place. Some form of tool or bowl is used to develop a dishlike depression in the sand, and the sides of the form are packed down firmly. A very dry mixture of concrete is prepared and placed very carefully into the sand mold. After drying, it is taken out of the sand and rinsed off and kept damp so that curing can continue. The stem is made in a similar manner; a cylindrical hole is made in the sand pile. A dry mixture of concrete is carefully placed into the hole and packed. Finally, the stem and mushroom top are mortared together.

Bill of Materials

1 pile	clean sharp sand for forming
1 qt	portland cement
3 qts	perlite (or pea gravel max. ¼ in. in diameter)
1 qt	sand for concrete
½ oz	color (if desired)

Sequence of Creation (Figure 15–3)

Preparation of the plan:

1. Determine the size of mushrooms that are to be made.
2. Determine their shape: rather flat and oval, evenly rounded, or highly domed. Should the stems be straight or curved (Figure 15–3a)?
3. Determine the finish: natural cement white, gray, colored with admixture, or painted after curing.

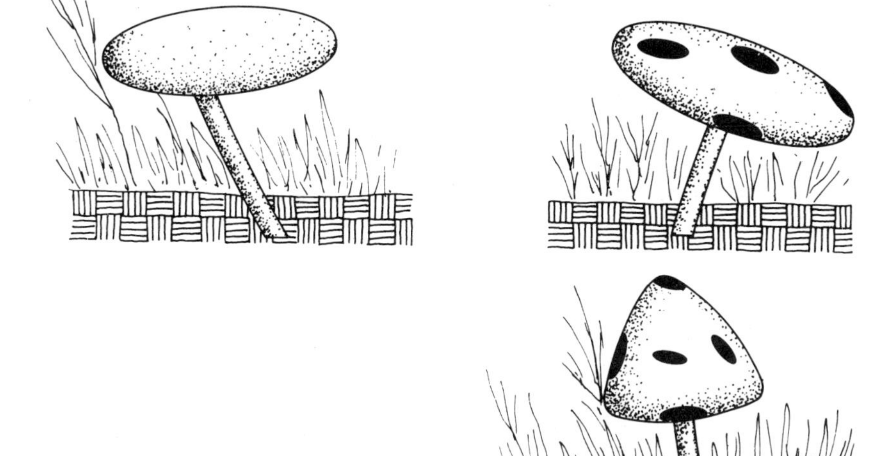

Figure 15–3a Sketches of Mushrooms

Preparation of the form, and pouring concrete for top and stem of mushroom:

1. Make the desired depression in the wetted sand (Figure 15–3b).
2. Mix a portion of dry mixed concrete. (*Note:* The mixture looks granular.)
3. Carefully place concrete into the molded sand and press firmly until the mold is full (Figure 15–3c).

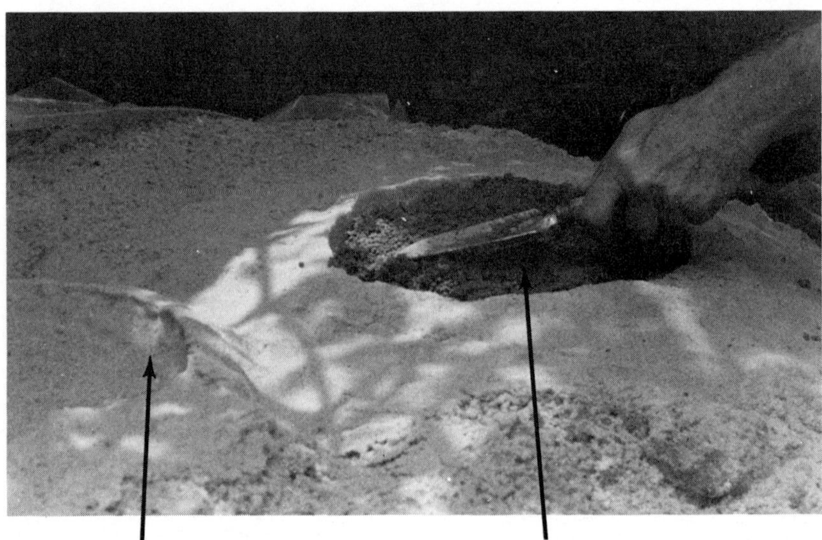

Figure 15–3b Molding Sand **Figure 15–3c** Filling Molds with Concrete

4. Allow to set for 24 hr and cure once removed from the mold and cleaned of excess sand.

Assembly and finishing:

1. Mix a small portion of mortar mix and sand, and mortar the top and stem (Figure 15–3d).
2. Coat the outer surfaces with color or concrete paint (Figure 15–3e).
3. Place in the garden by burying a portion of the stem in the soil (Figure 15–3f).

Figure 15–3d Mortaring Stem to Mushroom

Figure 15–3e Finishing Mushroom Tops

Figure 15–3f Finished Mushroom in Garden

PROJECT 15–4 GARDEN BENCH

Description

The garden or poolside bench is made of a variety of materials. The footing or foundation is made from concrete pavers purchased in builders' supply or nursery outlets. The vertical supports for the bench are readymade architectural cement blocks measuring 12 in. × 12 in. × 4 in. The seat is made from cedar or redwood and can be fashioned as a series of narrow pieces with alternate spacers or a solid flat board seat. Mortar is used to set pavers to the ground and 12-in. blocks to the pavers. The seat rests on top of the 12-in. block, and cleats under it keep it in place.

Bill of Materials

2	16 in. × 16 in. × 2 in. pavers
2	12 in. × 12 in. × 4 in. architectural decorator blocks
1	2 in. × 6 in. × 10 ft (makes a 16 in. × 36 in. bench)
1 cubic foot	mortar
6 medium shovelfuls	sand
2 medium shovelfuls	mortar mix
	tap water
1 qt	stain and sealer
½ lb	waterproof glue
½ lb	4d galvanized finishing nails
½ lb	6d galvanized finishing nails

Sequence of Creation (Figure 15–4)

Preparation of the plan:

1. Decide on the place for the bench, then determine its length.
2. Allow room to comfortably walk in front of the bench.
3. Decide on the finished height of the seat (should be approximately 15 in. to 16 in. from the ground).
4. Examine the subsoil to determine the need for gravel or other landfill.
5. Plan how the seat for the bench is to be made and finished.

Preparation of base and masonry unit installation:

1. Dig a hole several inches below grade level and, when approximately the size and shape of the pavers, fill with mortar. While wet but beginning to set, lay the pavers in place and level them (Figure 15–4a).
2. Make a bed of mortar for each architectural block and install the blocks according to the plan (Figure 15–4a). Make sure that these units are plumb and level, spaced properly, parallel to each other, and perpendicular to the front line of the pavers (Figure 15–4b).

(*Note:* On benches 6 ft or longer, use a center support made from the same materials.)

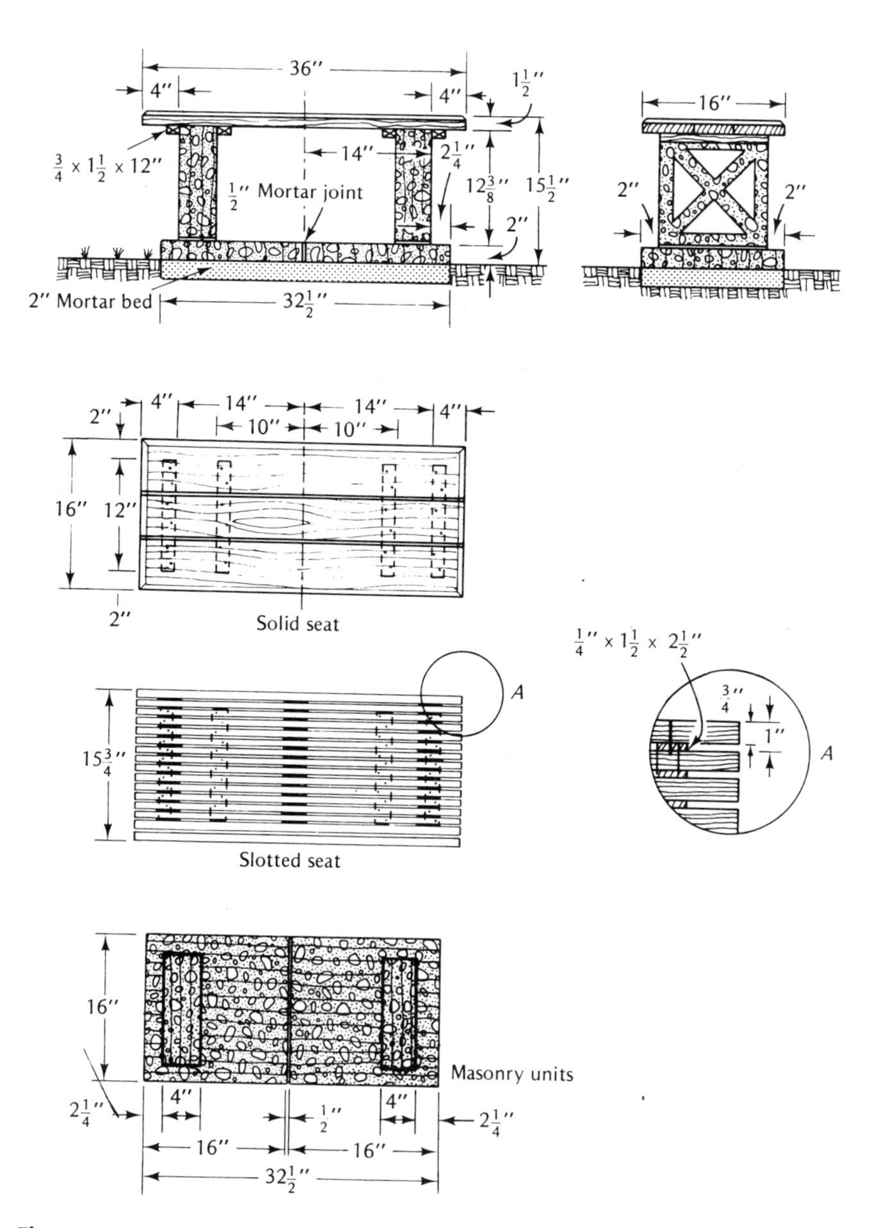

Figure 15–4a Dimensional Drawings for Garden Bench—Elevation Front and End Views, Seat Bench Construction Detail and Masonry Foundation (Top View)

Figure 15–4b Setting Masonry Units

Preparation of seat for bench:

1. *Solid seat:* a. Cut a 2 × 6 10 ft long into three 36-in. pieces.
   b. Cut four cleats 12 in. × ¾ in. × 1½ in. from the remaining 2 × 6 stock.
   c. Nail cleats to the underside of three pieces as shown in the plan.
   d. Bevel the upper edges and ends of the seat.
   e. Stain and seal.
2. *Slotted seat:* a. Rip a 2 × 6 into ¾ in. × 1½ in. strips.
   b. Cut sixteen 36-in. lengths from the strips.
   c. Cut four 12-in. lengths from one strip.
   d. Rip one strip to ¼ in. thick × 1½ in., then cut off 34 lengths of 2½ in. each.
   e. Alternately nail and glue a strip and block until all sixteen 36-in. pieces are used.
   f. Install the four cleats under the assembled seat.
   g. Bevel the edges and ends of the seat, stain, and seal.
3. Place the seat over architectural blocks (Figure 15–4c).

Figure 15–4c Completed Bench

PROJECT 15–5 GARDEN FISHPOND OR REFLECTING POOL

Description

A garden pool, whether used as a fishpond or reflecting pool, can be a very simple project. The one suggested in Figure 15–5a is the utmost of simplicity, yet has a certain formality. It is made from a variety of masonry units and concrete. Because it is so small, elaborate plumbing need not be included. After constructing the pool, all its surfaces should be plastered and then painted.

Bill of Materials

12	2 in. × 8 in. × 16 in. cap blocks
2	6-in. return or corner units (see Figure 15–5c)
10	6 in. × 8 in. × 16 in. blocks
1 bag	mortar mix
1 bag	cement
½ yd	sand
½ yd	gravel
1	overflow pipe, ½-in. copper tubing 8 in. long
1	4 ft × 6 ft piece 4 in. × 6 in. reinforcing wire for slab

Sequence of Creation (Figure 15–5)

Preparation of the plan:

1. Locate a position in the garden where the pool will be the focal point of interest.
2. Determine how much of the pool should remain above ground level (Figure 15–5a).
3. Determine if the pool size in the plan is too large or ill-proportioned.
4. Determine if the pool is too formal.
5. Plan on setting block while the concrete slab is setting up.
6. Plaster all surfaces, by brush, with a slurry mixture of mortar mix, sand, and water.
7. Plan to paint the pool interior blue.

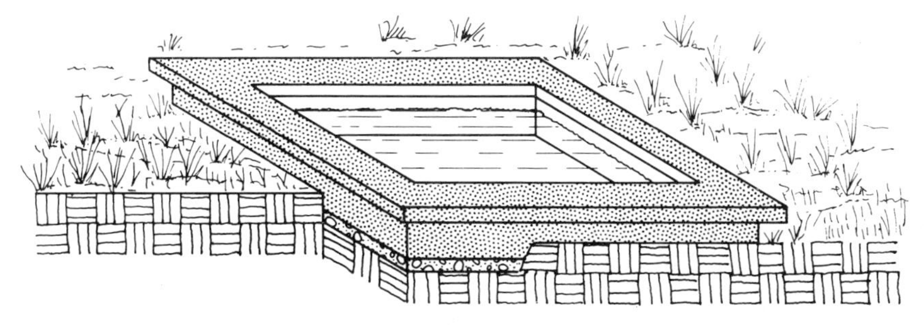

Figure 15–5a Garden Fishpond or Reflecting Pool

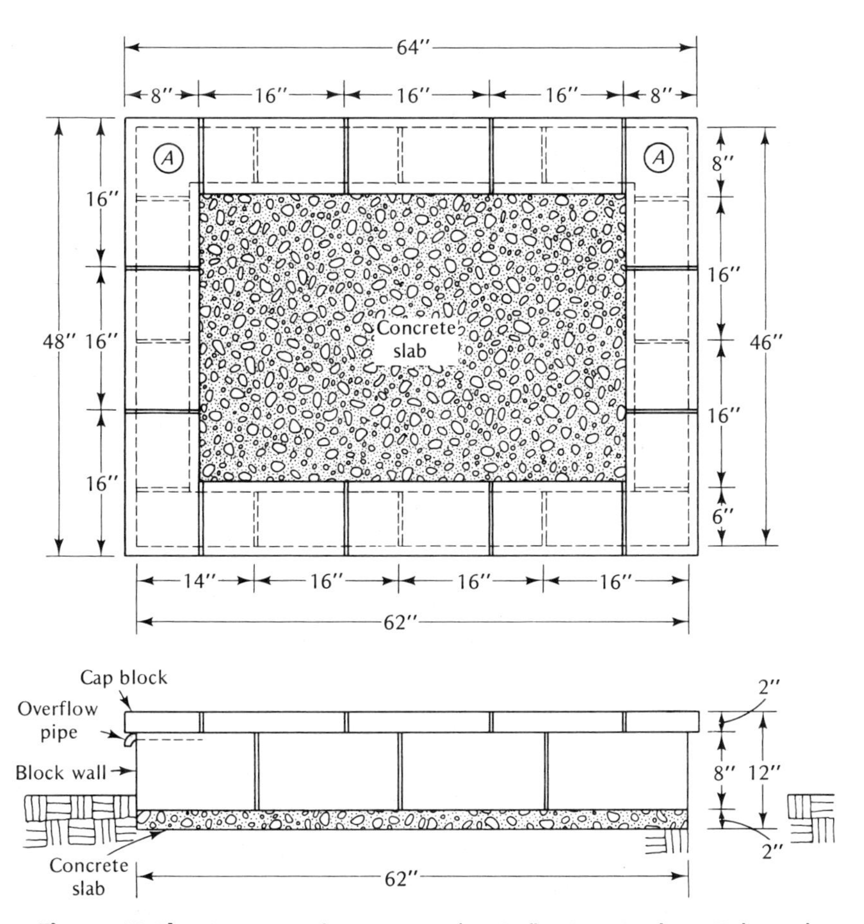

Figure 15–5b Dimensional Drawings for Reflecting Pool or Fishpond

Preparation of the base slab, erection of pool sides, and cap:

1. Remove the earth to the desired level in an area 4 ft wide by 5 ft 4 in. long. Try to keep the sides of the earth straight to avoid using wood forms.
2. Check the dugout area for level in several directions.
3. Mix a batch of concrete in a rented electric concrete mixer. Use a 1:3:4 mixture.
4. Place the wire into the formed area and fill the formed area with 2 in. of concrete. Screed even and float when the surface has lost its sheen.
5. Set the blocks into fresh concrete after floating. Be sure to follow the schedule shown in Figure 15–5b. [*Note:* Two corner blocks must be cut to 14 in.; the rest are all full length (Figure 15–5c).]
6. Mortar all the joints between the blocks and make sure that all are aligned and that all corners are at 90 degrees.

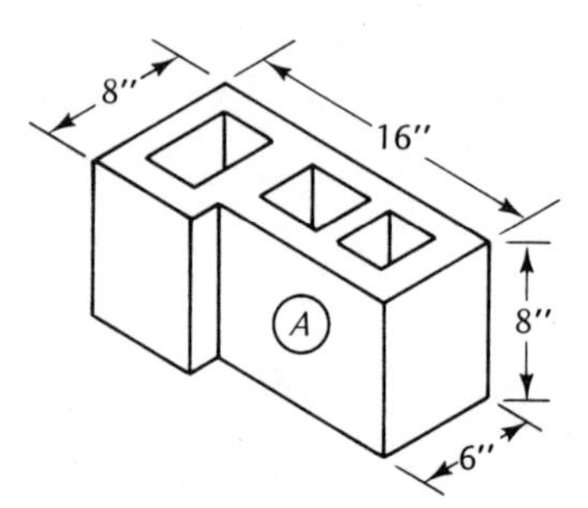

Figure 15–5c Corner Block

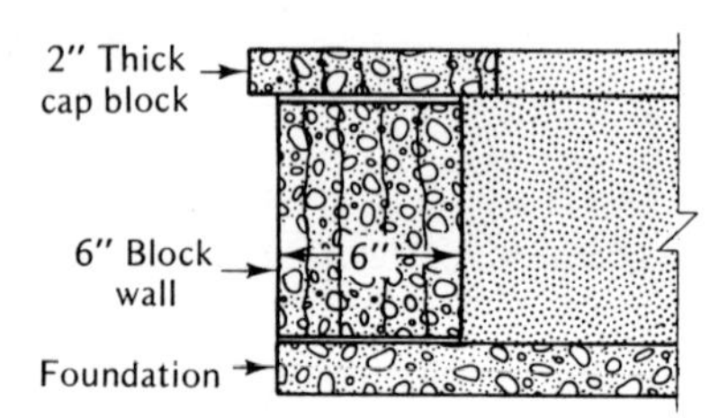

Figure 15–5d Section Detail Showing Cap Block Overlap over Wall Block

7. Mortar the 8-in. cap blocks in place; be sure to overlap the 6-in. wall blocks 1 in. each side (Figure 15–5d). (*Note:* Insert the overflow pipe in a mortar joint just under a cap block.)

Finishing:

1. Mix a slurry of mortar, color if desired, and water and apply with a wide paintbrush to all surfaces of the pool. Let dry.
2. Paint the interior of the pool surface up to the water level.

Section IV

Maintenance and Repair of Concrete and Masonry

This section includes four chapters that identify maintenance and repair procedures used successfully on concrete and masonry. Each chapter follows the previous chapter organization. The stated objective should be clear to the reader. The description following the objective is written and illustrated for clear understanding.

The chapters include in order: repairing cracks in concrete and masonry, waterproofing interior walls, repairing steps of concrete and brick, and removing stains and cleaning masonry surfaces.

Chapter 16

Repairing Cracks in Concrete and Masonry

Admixture finely ground minerals that are used to color mortar or concrete (may be other materials, see Chapter 1).
Feathered edges any edge in a surface that resembles the thin edge of a bird's feather. Two adjacent surfaces having an angle of several degrees.
Screeding the leveling or evening off of a concrete surface with a board or screed.
Tuck pointing the filling in with fresh mortar of a cutout or defective mortar joint in masonry.

OBJECTIVE—INTRODUCTION

The natural elements are always present—with their pressure, force, weight, and pull; and even though concrete, mortar, and masonry units have some elasticity, a point is reached where a crack develops. Then, too, there is the human error of bad engineering, shoddy workmanship, or the builder who has not followed plan specifications. Cracks resulting from whatever reason must be repaired, and this identifies the objective of this chapter: *to be able to select the correct method and materials, and to repair a crack in either concrete or masonry structures.*

Since both concrete and masonry structures are included in the objective and their repair procedures differ somewhat, each is examined separately.

REPAIRING CRACKS IN CONCRETE

Because there is seldom a pattern in the forming of the concrete mass, cracks tend to change direction randomly. As a rule, they follow the line of force that causes the crack. These may be circular, zigzag, rather straight, or on a bias. Usually, though, the repair process is simplified by all irregularities being centered within a repair area, as Figure 16–1 shows.

Observing the figure closely, one sees that some concrete is designated for removal either side of the crack. This is necessary so that the repair is effective and lasting.

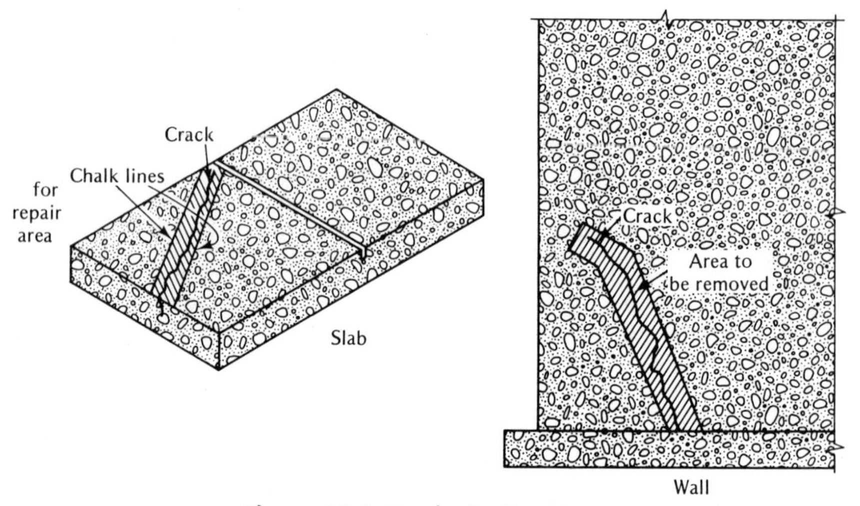

Figure 16–1 Cracks in Concrete

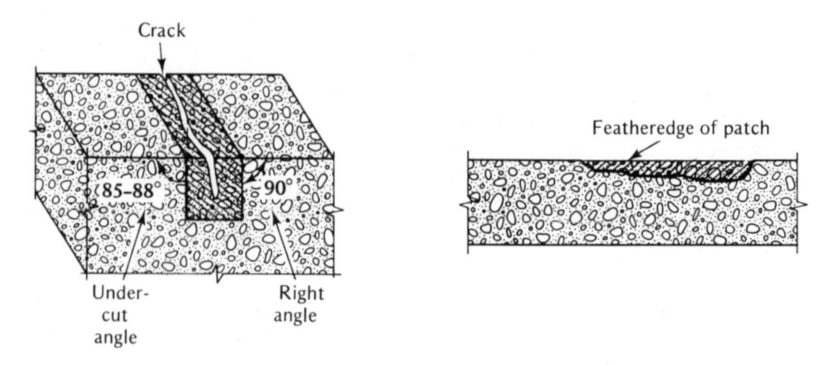

Figure 16–2 Cutaway of Cracked Concrete

Figure 16–2 provides more detail about the fundamental requirements for proper repair. Notice that the sides of the area to be patched are at right angles to the surface or cut slightly under. Feathered edges, those whose surface are almost parallel to the concrete's surface, do not allow for adequate thickness of patch and are subject to erosion and chipping.

The crack may be short, in the range of 1 or 2 ft, or quite long. It may be in a wall or in a driveway or slab. If small, hand tools can be used to prepare the area for the patching concrete or mortar. If the crack is quite long, a power saw fitted with a *diamond* or *carborundum* saw blade can be used to cut the edges into the concrete.

The method of repair follows the general sequence of steps:

1. Outline the area of concrete to be removed. This can be done easily with a chalk line. Each line should be snapped several inches back from the crack.
2. A cold chisel or brick chisel and hammer should be used to clean out the concrete within the chalked area. The depth of the cleaned hole should be at least *1 in.* deep, or deeper if the crack goes deeper. The sides of the hole must be at right angles or slightly undercut. (Refer to Figure 16–2.)
3. All chiseled areas should be blown clean and then the area must be wetted for several hours before filling.
4. Grout, made from cement and water, should be brushed in the area to be patched. This makes for a better bond.
5. Mortar or concrete should be made fairly dry to minimize shrinkage, and placed in the hole in small quantities. As each small batch is placed, it should be well packed and roughened so that good bonding results.
6. Next, the final batch of mixture should be screeded slightly above the old concrete's surface. When it is ready for floating and troweling, these operations will make it flush with the surrounding surfaces.
7. Cover the patch if undesirable curing conditions are present.

REPAIRING CRACKS IN WALLS MADE OF MASONRY UNITS

Seldom, if ever, does a cement block, concrete block, brick, or tile crack in a wall. All cracks seem to occur in the mortar joints. When they do, they usually follow a stair-step pattern.

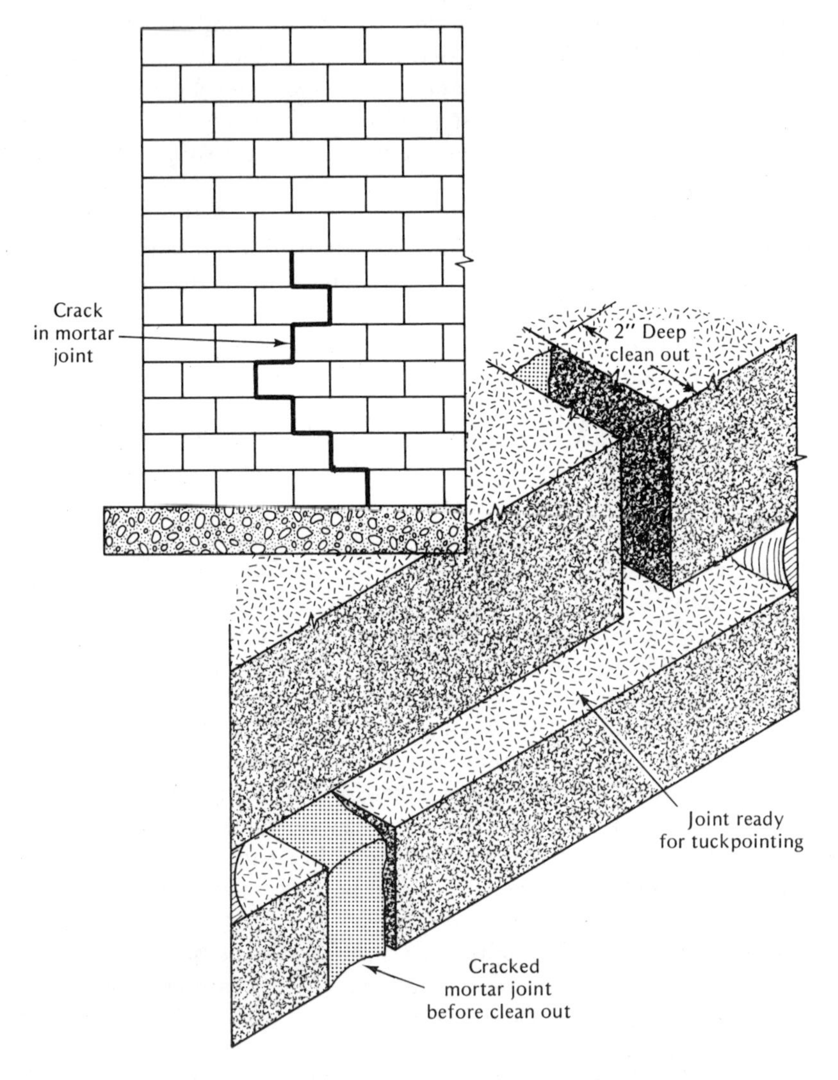

Figure 16–3 Repairing Cracks in Walls Made of Masonry Units

The fault of the rupture was defined earlier as natural, engineering, or poor workmanship, but since the building has been erected, there is little that can be done to correct the basic problem. However, the cracks can be repaired quite simply, and the following description is provided and illustrated in Figure 16–3.

1. Chip or chisel out mortar between masonry units to a depth of 2 to 2½ inches. Make sure that surfaces of the masonry units are almost clean of old mortar.
2. Blow all joints clean with air.
3. Wet the cleaned-out area by using a fine water spray or brush and water. This is important if a good bond is to be made with the new mortar.
4. Prepare a small batch of mortar and keep the mixture fairly stiff.
5. Using a pointing trowel, tuck-point small quantities of mortar into the joint; pack firmly. Repeat this step time and again until the joint is full of mortar.
6. Strike the joint to make it look like all the others on the wall. If a stiff mortar is used, the operation may be done immediately. If a less-than-stiff mortar was used, wait 20 to 30 minutes before striking the joint.

(*Note:* Mortar should be colored with an admixture if old mortar was colored.)

A final word on the problem of cracks. If a crack or several cracks appear in a structure, postpone repairs until the full extent of the cracked area has developed. If the repair is started too early, more work may be involved because the repairs may also crack. Cracks can be marked with a chisel so that when weekly or monthly inspections are made, changes will show. If the cracks occurred because of a serious change in the natural environment—earthquake, flood, subsoil erosion, or other—an engineer or architect should be called to examine the structure for soundness. He should determine if a major shoring or reinforcement job must be performed to avoid structural collapse.

Chapter 17

Waterproofing Interior Walls

Bituminous coating or mastic asphalt or coal-tar pitch in a heavy cream or light paste consistency, used to seal out water.

Efflorescence a powder or stain deposit of water-soluble salts found on the surface of masonry units, bricks, tile, etc.

Parging the act of plastering a wall with thin coats of mortar, usually on the inner surface of the wythe.

Plastering the act of applying thin coats of mortar mixtures to the outer surface of a masonry wall.

OBJECTIVE—INTRODUCTION

The most prevalent reason for waterproofing a wall is because of poor or shoddy workmanship. Poorly made mortar joints, improperly prepared bricks or blocks, and inadequate quantities of mortar in each joint all create conditions for leaks. Another reason for waterproofing is poor bonding where hairline cracks form and allow water to seep into the wall. Once started, erosion continues and finally breakthrough occurs. Unfortunately, the leak may not occur until several months or years after the wall is completed.

Even though it is possible that neither of the causes were present during construction, walls in continuous contact with earth are subject to groundwater and dampness; and a point is eventually reached where the moisture penetrates the block, brick, or mortar. In many cases moisture even penetrates concrete walls. This then identifies the objective for this chapter: *to be able to waterproof a wall to eliminate penetration of moisture.*

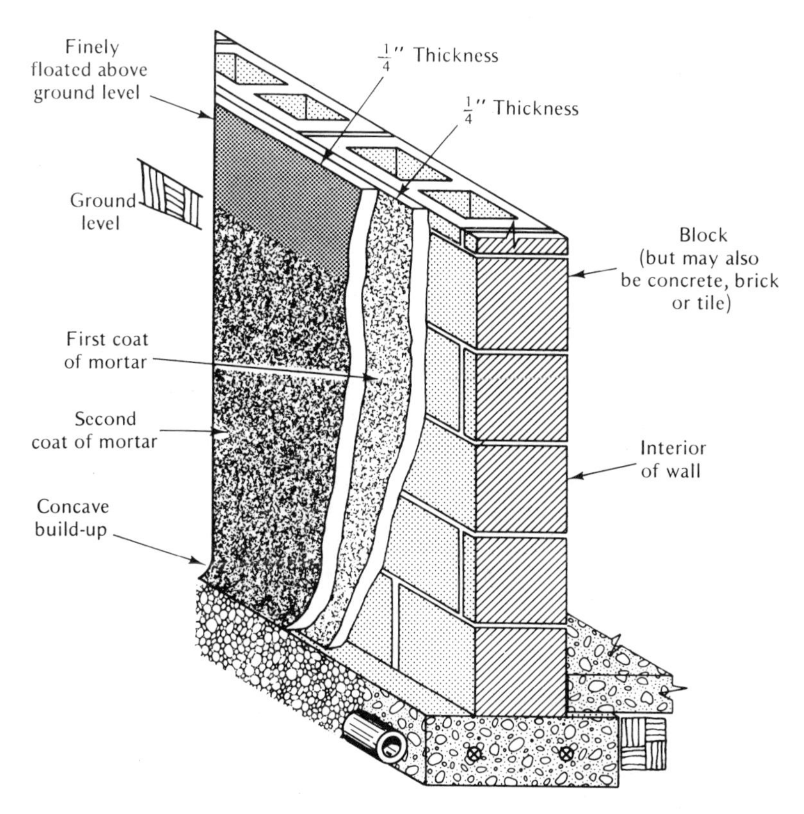

Figure 17–1 Plastering a Wall for Watertightness

PLASTERING AND PARGING METHOD

The most common method of waterproofing a wall using a rich cement mortar mixture is shown in Figure 17–1. The mixture is plastered to the clean outer surface of the wall after the wall has been sprayed lightly with water and allowed to absorb the water. This slight buildup of moisture prevents absorption of water from the mortar. Recall that water must be present if hydration is to continue. The process is as follows.

1. Mortar is scooped onto a steel trowel and then applied to the wall in an upward motion. The thickness should be made ¼ to ⅜ in. This task is repeated until the entire wall is coated. A lightweight screen mesh is usually used to score the first

coat before it sets. This makes for a more secure bond with the second coat.

2. A concave buildup of mortar is formed at the footing–wall joint. This buildup will cause water to shed away from the base rather than accumulate and penetrate.
3. A second coat of mortar, also ¼ in. thick, is applied over the first coat after the first one has dried. The second coat may be rough floated or finely floated or steel troweled. Since it is unlikely that the plaster will show below ground level, little effort is spent on appearance; it is complete coverage that is important. However, exposed surfaces should be floated smooth.

Back-parging, in contrast to plastering, is done on the inside of a wythe. But for it to be effective, the mortar from joints must not protrude beyond the masonry unit. Then too, parging is frequently performed concurrently with wall erection. The mortar for parging must be rich in cement and applied at least ⅜ in. thick. It may be applied with a pointed trowel or a steel trowel and it must completely cover the back surface of the masonry wall.

WATERPROOF COATINGS METHOD

Waterproof coatings may be bituminous, specially prepared waterproof paints, or any of the various portland cement paints on the market. For most of these coatings, no plastering is absolutely required, but in many locales plastering is done prior to bituminous application. The various coatings listed above require different methods of application, as follows:

1. The bituminous coating method shown in Figure 17–2 uses tar, pitch, or asphalt in a creamy to slightly pasty consistency and a brush, broom, or mop. It is rubbed into the surface as it is applied so that all pores in the wall's masonry units, concrete, and mortar are filled. Usually two coats are applied, with a slight delay between applications.
2. The waterproof paint method employs the application of transparent waterproofing paint. The paint is applied by brush or spray gun to surfaces above ground level. Oil base paint and some of the newer acrylic paints in colors may also be used. Manufacturers' instructions should be followed for successful application.

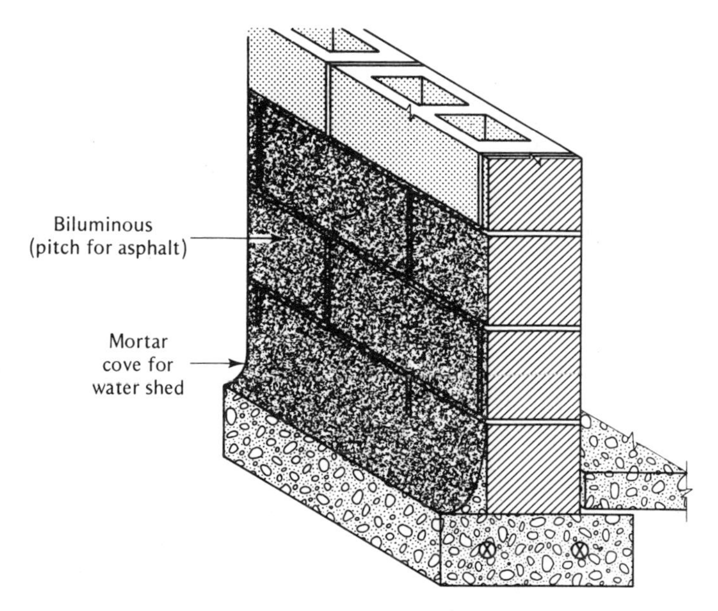

Figure 17–2 Bituminous Coating on Block Wall Below Ground Level

3. The portland cement paint is specially prepared to be applied to surfaces above ground level. These paints are frequently sold in powder form and mixed with water prior to use, and during the painting the paint should be stirred frequently to keep the powder in suspension. Certain preliminary steps must be followed if a lasting quality is expected. For example, the surface to be painted must have been completed at least 1 month, and it must be entirely free of efflorescence and foreign matter. Just prior to brushing it on the surface with old-fashioned whitewash brushes, the wall must be sprayed with water. Applying the paint with a sprayer does not produce the water tightness that brushing it on does. Further, the paint is usually applied to mortar joints first and with great care, so that all pinholes or hairline cracks are filled; then the remaining surface is coated. After a waiting period of at least 12 hr, the wall is once again sprayed with water and a second coat of paint is applied.

Chapter 18

Cleaning Concrete and Masonry Surfaces

Ammonium chloride saltlike substance available in most drugstores.

Lime-free glycerine a liquid form of glycerine frequently found in hand lotion, used to remove rust stains.

Muriatic acid hydrochloric acid, one of the most widely used inorganic cleaning agents for concrete and masonry.

Sodium citrate large saltlike crystals used in cleaning rust.

Talc finely ground powder from rock, used as part of the paste in cleaning.

Trichloroethylene a highly refined solvent (cleaning fluid); may be toxic and dangerous if breathed for long periods.

Whiting a powdered chalk used frequently in paints and available at paint stores.

OBJECTIVE—INTRODUCTION

Many substances exist that cause staining on concrete and masonry surfaces. Some are deposits of earth or windblown organic vegetation that are easily washed off with a common household bleach used in full strength followed by a clean water rinse from a garden hose. Other stains are not so easily removed, and the compounds and chemicals used may cause harm if improperly used. These concerns then identify the objective for this chapter as: *to be able to properly clean concrete and masonry surfaces and to protect them from deterioration because of cleaning agents.*

CLEANING DRIED MORTAR FROM BRICK OR BLOCK

Some mortar invariably is spilled, splashed, or dropped on the surface of freshly laid masonry. Although it would be easiest to clean these deposits while they are still fresh, any cleaning done would also disturb the mortar between the masonry units and thus the entire wall or pavement. Because of this, mortar deposits are cleaned after the mortar has *set.* Masons usually call this cleaning activity a wash down.

Before describing the process of cleaning, it is extremely important to understand that if acid is used to remove mortar deposits, the acid can also dissolve mortar in the joints. In addition, acids may adversely affect the surface and texture of brick, tile, or block. Therefore, acids, if and when used, must be thoroughly rinsed from the surface to which they were applied.

The process of cleaning dried mortar is as follows:

1. While the wall is dry, chip, scrape, and rub as much of the dried mortar from the surface as possible. Large pieces should be chipped away with a chisel; smaller pieces and deposits may be removed with a stiff-bristle wire brush or steel wool.
2. Presoak the wall with clean water. This loosens mortar freed in step 1 and washes mortar and dirt from the surface.
3. For light deposits of mortar, try a scrub down using ½ cup of of dishwasher detergent (trisodium phosphate) in 1 gal of water. A stiff bristle brush must be used in the scrubbing operation.
4. For stubborn and heavier deposits of mortar, a scrub down using muriatic acid in 1 part to 9 parts of water mixed in a *nonmetallic* pail is used. Once again a stiff bristle brush is used. [*Caution:* Since muriatic acid is dangerous to use, (1) *pour the acid into the water,* not the water into the acid, and (2) wear protective eye goggles, gloves, long-sleeved shirt buttoned to the collar and at the sleeves, and a hat. If available, a plastic face mask is even better than just eye goggles, since the entire face is protected.]
5. Do not allow the acid solution to dry on the surface or run down on the dry wall surfaces. *Keep the wall wet.*
6. Wash the acid solution from the surface by spraying the surface with generous quantities of water, and brush the surface vigorously to dislodge all acid that may have penetrated the mortar or surface material of the masonry units.

REMOVING EFFLORESCENCE

The white powdery substance found on masonry walls is called efflorescence. It generally consists of one or more water-soluble salts that were present in the materials used to make the brick or tile. These salt deposits usually surface as water is evaporated from the brick or tile. Fortunately, it is not difficult to clean off.

1. Presoak the surface thoroughly with water and brush the surface vigorously.
2. Mix 1 part muriatic acid with 9 parts water in a nonmetallic pail.
3. Wear protective clothing and, with a stiff bristle brush, rub the acid solution into the surface.
4. Wash the surface thoroughly with water after applying the acid solution. It is absolutely essential to remove all acid to avoid deterioration to mortar and brick or tile.

The following descriptions apply to concrete surfaces as well as to masonry. Stains that are listed in this section are not usually caused by or during the construction phase of the job. They occur because of the use of the surface, such as a driveway, or because of their proximity to metals and oxides formed from the metals, or by the vegetation that comes in contact with the surface.

METAL OXIDE STAINS

The proper method to use in cleaning oxide from a surface is to make a paste of chemicals in talc or whiting powder and apply the paste to the stain. The paste is left to dry, and during its drying it neutralizes the oxide and draws it from the surface into the talc. Table 18–1 provides formulas for cleaning iron oxide and copper oxide.

ORGANIC STAINS

Organic stains result as a rule from grass, paper, straw, plant growth, and dirt. The surfaces should be thoroughly washed, cleaning agent applied, and rinsed. Scrubbing is essential to a successful job. Table 18–2 provides formulas for cleaning several organic stains.

TABLE 18–1 REMOVING OXIDES FROM MASONRY UNITS AND CONCRETE

Stain	*Cleaning agent*
Iron oxide (rust)	Paste: 1 part lime-free glycerine, 1 part sodium citrate, 6 parts lukewarm water, mixed with talc or whiting
Copper oxide	Paste: 1 part ammonium chloride, 4 parts powdered talc

TABLE 18–2 REMOVING ORGANIC STAINS FROM MASONRY UNITS AND CONCRETE

Stain	*Cleaning agent*
Plant growth and grass	(1) Ammonium sulfamate (certain weed killers have a suitable substitute if this compound is not available) (2) Household bleach in full strength [very effective, in removing (black) spores]; allow to dry, then rinse
Paper, straw	Household bleach, full strength; allow to dry, then rinse
Oil and tar	Kerosene, naphtha, trichloroethylene or benzene (*Caution:* Toxic fumes are given off; follow chemical operation with a thorough water rinse)

USING THE CAUTIOUS APPROACH

Any cleaning task involving chemicals and compounds should be approached with caution. First, the type of stain should be identified. Then a small portion of the cleaning agent should be made up and applied to a small out-of-the-way part of the stained surface. The technique to be used should be followed step by step in this small trial area. During each phase of the job, all changes, good and bad, should be noted and written down. The time for each phase should also be recorded, especially the part where acid is on the surface, and how long it takes to wash off the acid solution. Then when the surface has dried, an examination must be carefully made to determine if the surface has been cleaned and if there has been any deterioration, change in color, presence of residue, or anything else that looks unnatural.

If anything unusual occurs ask someone more qualified for assistance and direction. It may also be true that what looked like one type of stain is actually another.

Chapter 19

Repairing Steps of Concrete and Brick

Bond the uniting of two materials, such as by gluing.
Grout cement and water (usually including sand) for use in this chapter on repairs without sand.
Mortar a mixture of portland cement, sand, and water with additives; also made from masonry cement, sand, and water.
Riser the vertical distance between treads and the back surface of this distance.
Tread the portion of a step that one walks on.
Undercut any surface that is tapered back from its top surface more than perpendicular.

OBJECTIVE—INTRODUCTION

It is very rare for a set of stairs to totally collapse, but it is not at all rare to have one or more treads of a stair crumble, crack, and fall off or have one or more bricks dislodge. Penetration of moisture into the concrete or mortar joint results in erosion, expansion and contraction, and further decay. Improperly prepared mixtures of either concrete or mortar are frequently causes of stair decay. These causes and their results are the objective of this chapter, which is: *to be able to repair a set of steps made from concrete or brick.* Since the one objective identifies a single purpose but the work could be accomplished on either a set of concrete steps or a set of brick steps, each type is examined separately. It will be shown that different tasks are needed for each kind of stair.

REPAIRING A CONCRETE SET OF STEPS

Figure 19–1 illustrates a typical and rather severe problem of where one of the step treads has broken. The remaining surfaces are, as expected, very irregular and, of course, this area becomes contaminated very quickly with dirt, sand, and vegetation. Recall from Chapter 8 that there are no reinforcement bars in the tread-riser area. All bars are located below. On the one hand, this means that cutting or chiseling away additional concrete is simplified, but on the other hand, there is no reinforcement for the patching concrete to bond to.

Two methods of repair are illustrated and explained. Circumstances existing in individual cases should dictate which method is best suited. Obviously, the more severe the break, the more the need for reinforcement steel.

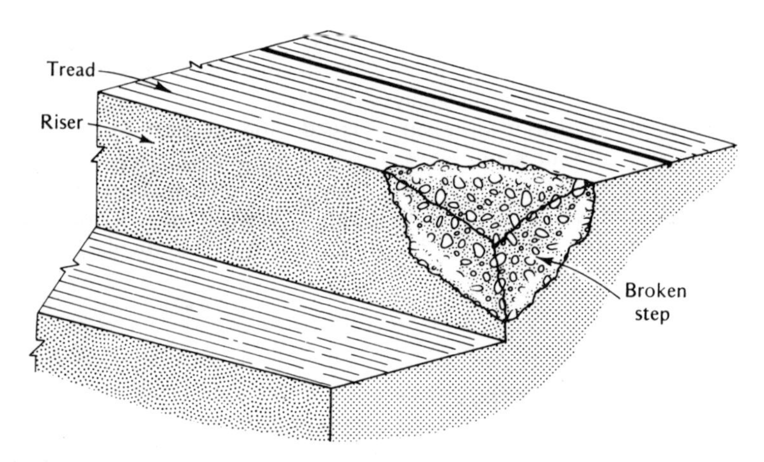

Figure 19–1 Concrete Step with a Broken Step Tread

Repairs Without Reinforcement Rods

Figure 19–2 should be used as a visual aid to understanding the process outlined below. Most important is to prepare anchoring probes or holes into the remaining step as well as to make sure that all decayed concrete has been removed.

Procedure:

1. Chip away all concrete near the break that has or appears to

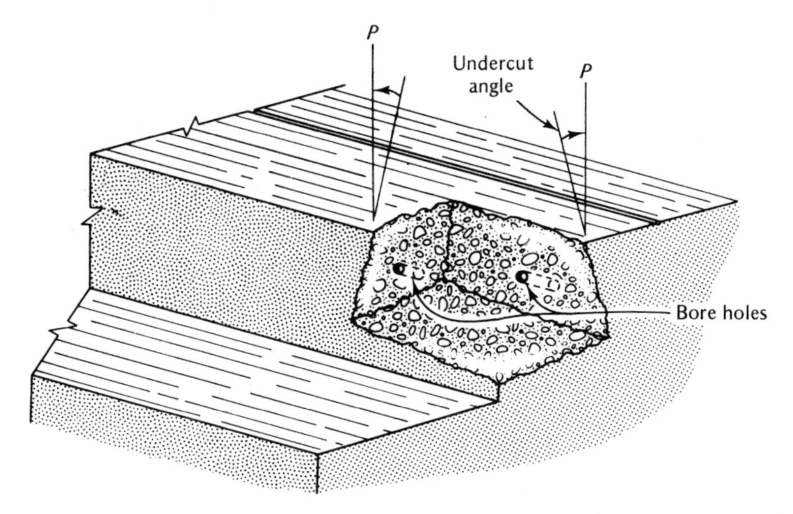

Figure 19–2 Repairing a Concrete Step Without Reinforcement Rods

have decay. Only clean surfaces of cement, sand, and aggregate should show.

2. Chisel holes into the cleaned area with ⅝-in. or ¾-in. star drills or electric drills using masonry bits. Make these holes at different angles and about 2 in. deep.
3. Undercut the edges of the break with chisels to assure a secure bond of patch. Blow the area clean of all materials.
4. Apply a wood form, if necessary.
5. Mix a small batch of portland cement and water to form a paste, and brush this mixture on all surfaces of the cleaned-out area. Do not fill bore holes with cement.
6. Mix a batch of concrete using cement, sand, pea gravel, and water to a fairly stiff consistency and fill the area. Force concrete into the holes, make sure all crevices are securely packed, and screed the concrete slightly higher than the surrounding hardened concrete.
7. When the concrete has lost its sheen, float the surface flush with the surrounding concrete, matching the texture of the other steps.
8. When the concrete has set, remove the form, *or* if the form is the riser, remove it after completing step 7 and finish as in step 7.

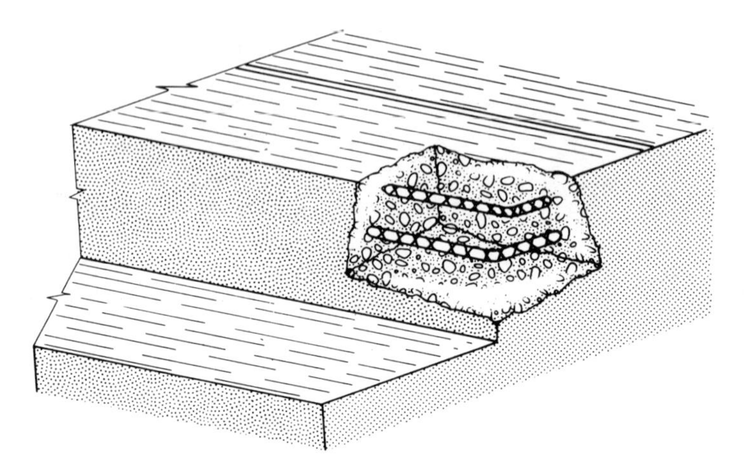

Figure 19–3 Repairing Concrete Steps Using Rods

Repairs with Reinforcement Rods

Figure 19–3 should be used as a visual aid to understanding the process outlined below. The main difference between the two methods is the inclusion of rods instead of bore holes. Holes, of course, must be bored into the concrete so that the rods can be anchored.

Procedure:

1. Chip or chisel away all deteriorated concrete around the break. Make surfaces near the surface perpendicular to the surface or slightly undercut.
2. Decide where to install either one or two ⅜-in. or ¼-in. pieces of rod, then drill pilot holes at least 2 in. deep and at least ½ in. larger than rod diameter.
3. Blow out the chiseled area and holes.
4. Make a grout of cement and water and brush this into the cleaned area.
5. Install the steel rods and force grout into the holes where the rods are. Brace the rods, if necessary.
6. Install the wood form as required.
7. Mix a batch of concrete to a rather stiff consistency. Use a pointing trowel to place it into the area. Pack the concrete well into all crevices and around the rods. Remove the braces on the rods, if used, as each rod is encapsulated with concrete.

8. Screed the concrete slightly above the finished surface of the remaining step tread.
9. When the concrete has lost its sheen, remove the form, float the surfaces, and finish to the texture of the other steps.

REPAIRING A BRICK SET OF STEPS

Repairing a set of steps made of brick is much simpler than repairing concrete steps. A single set of guide lines is useful for replacing corner

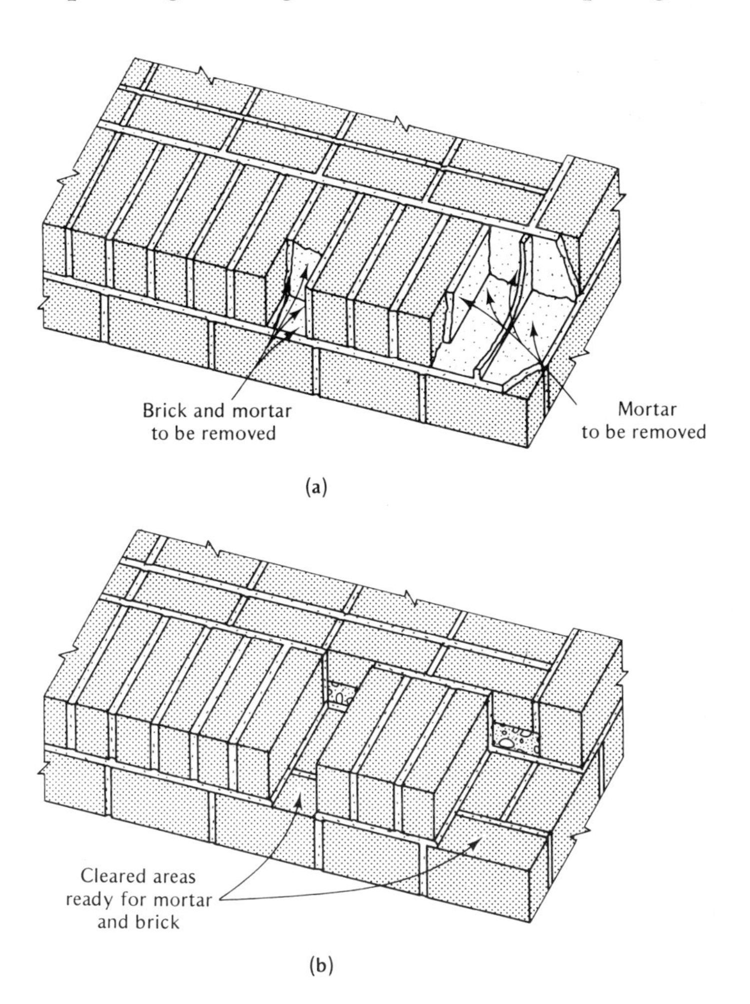

Figure 19–4 Repairing a Set of Brick Steps

or intermediate bricks. Generally, cleaning of old mortar must be thorough, then proper wetting must be performed, followed by brick installation and joint striking. Figure 19–4 shows the process outlined below.

Procedure:

1. Remove loose bricks and those that appear to have decayed mortar in their joints.
2. Chisel all old mortar from surrounding surface, below, alongside and behind. Blow the area clean to make periodic inspections. Clean the bricks as well.
3. Wet the cleaned area with water, soaking it properly. Also wet the bricks. This step is essential to prevent the mortar's moisture from being drawn into the bricks.
4. Mix a rich batch of mortar (white mortar mix is usually better for matching colors than gray is).
5. Place a generous amount of mortar into the area, base, sides, and back, then install one brick. Butter the side of the second brick and install it.
6. Trowel off the excess mortar.
7. When the mortar has begun to set (30 to 45 minutes), strike the joints. Even if they are to remain flush, strike them and tuckpoint to ensure a sound, watertight surface.

References

Brick Floors and Pavements, Brick Institute of America.
Brick in Landscape Architecture, Terraces and Walks, Brick Institute of America, 1976.
Builder 3 and 2, Bureau of Naval Personnel, 1970.
Cement Mason's Guide to Building Concrete Walks, Drives, Patios and Steps, Portland Cement Association, 1973.
Concrete and Masonry, Department of the Army, 1970.
Concrete Improvements Around the Home, Portland Cement Association, 1968.
Concrete Masonry Handbook, Portland Cement Association, 1976.
Design and Control of Concrete Mixtures, Portland Cement Association, 1968.
Estimating Brick Masonry, Brick Institute of America.
Fireplaces and Chimneys, Department of Agriculture, 1971.
How Big Is a Brick? Brick Institute of America, 1975.
Masonry Design Manual, Masonry Industry Advancement Committee, 1972.
Masonry Veneer, Masonry Institute of America, 1974.
Patterned Finishes for Slabs, Portland Cement Association, 1973.
Plywood for Concrete Forming, American Plywood Association.
Portland Cement-Lime Mortars for Brick Masonry, Brick Institute of America, 1972.
Principles of Clay Masonry Construction, Brick Institute of America, 1973.
Recommended Practice for Concrete Form Work, American Concrete Institute, 1968.
Recommended Practice for Engineered Brick Masonry, Brick Institute of America, 1975.

Recommended Practices for Laying Concrete Block, Portland Cement Association.

Small Retaining Walls, Portland Cement Association.

The Structure of Building Specifications, Department of Commerce, 1976.

Uniform Building Code, International Conference of Building Officials, 1973.

Appendix A

All pictures of tools are by courtesy of the Goldblatt (G), Marshalltown (M), and Stanley (S) tool companies.

Chalk line (G)

Groover (G)

Edge tools (G)

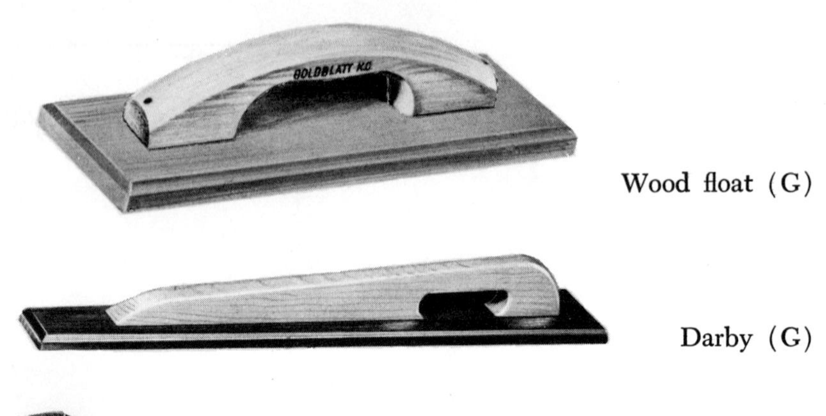

Wood float (G)

Darby (G)

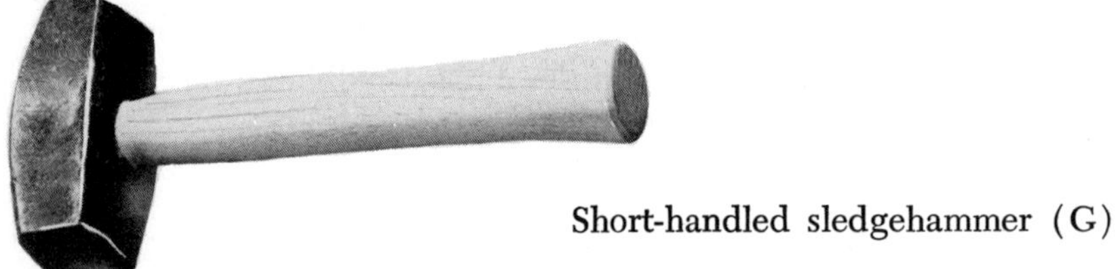

Short-handled sledgehammer (G)

Mason's line (G)

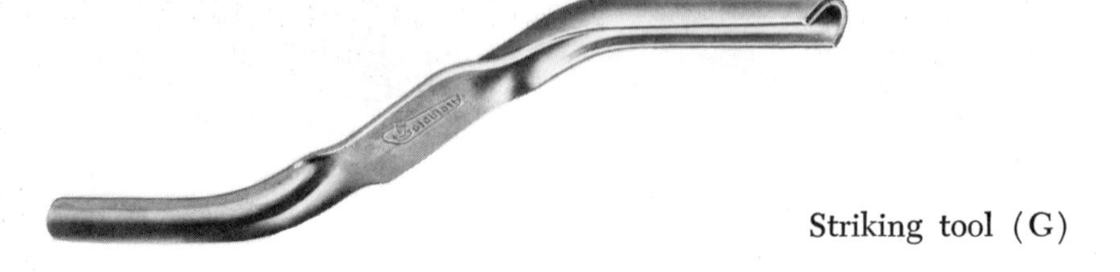

Striking tool (G)

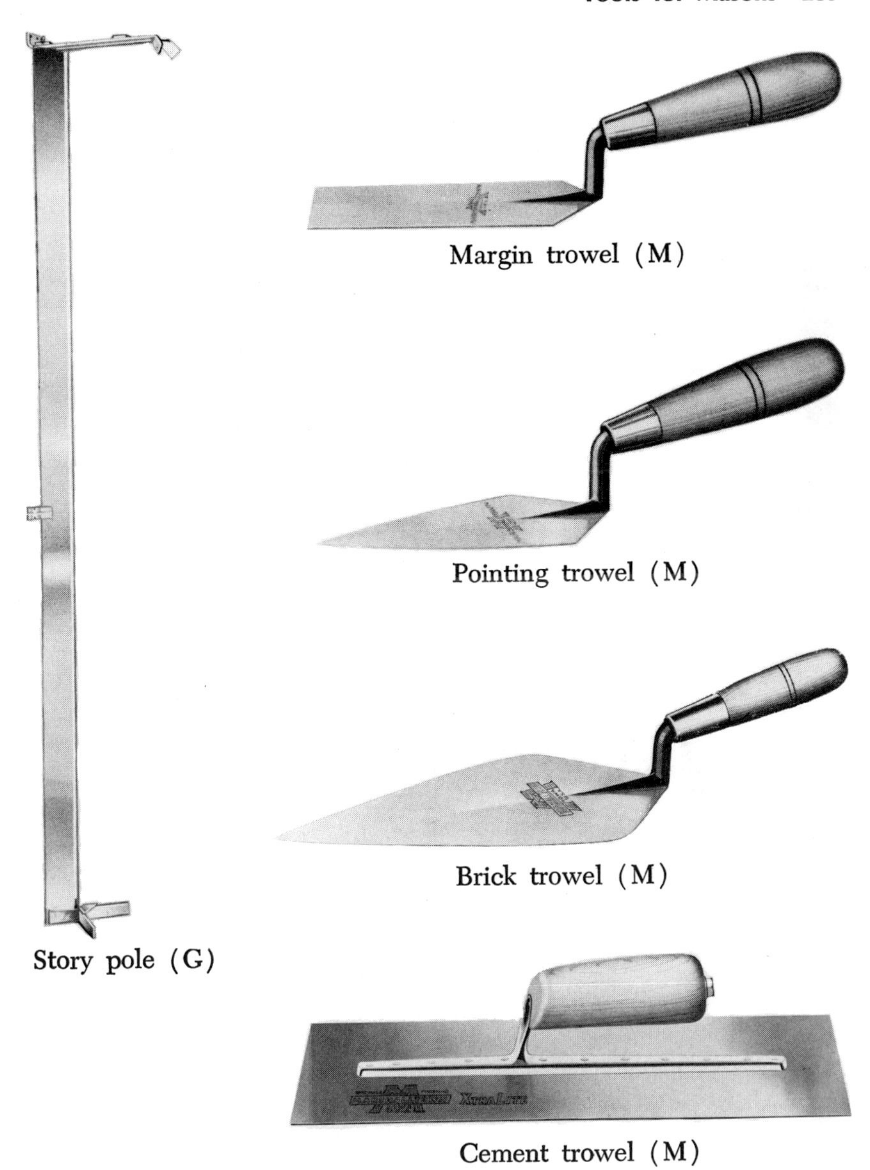

Margin trowel (M)

Pointing trowel (M)

Brick trowel (M)

Story pole (G)

Cement trowel (M)

Line level (S)

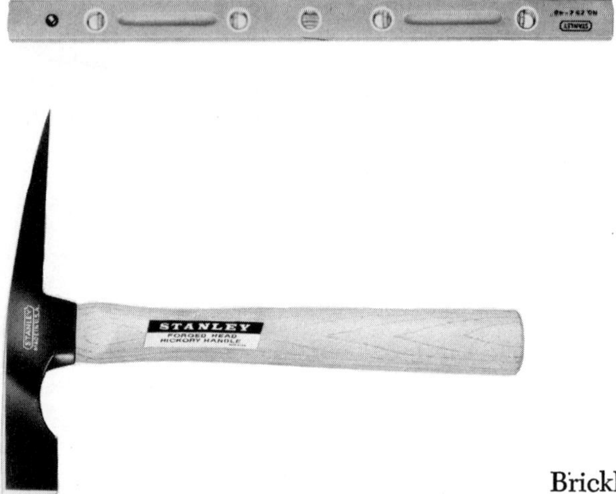

Bricklayer's hammer (S)

Half-hatcher (S)

Level vials (S)

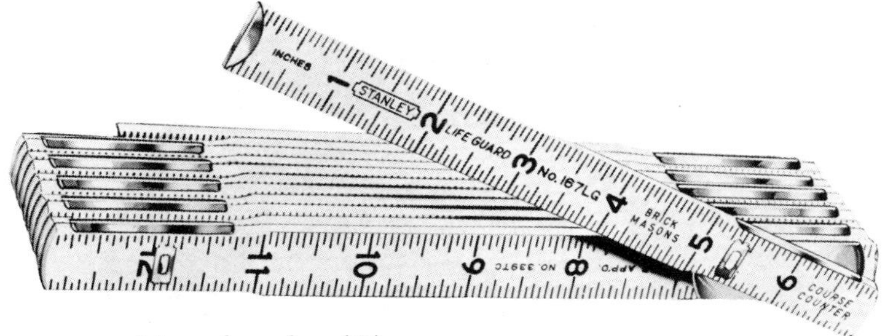

Mason's ruler (S)

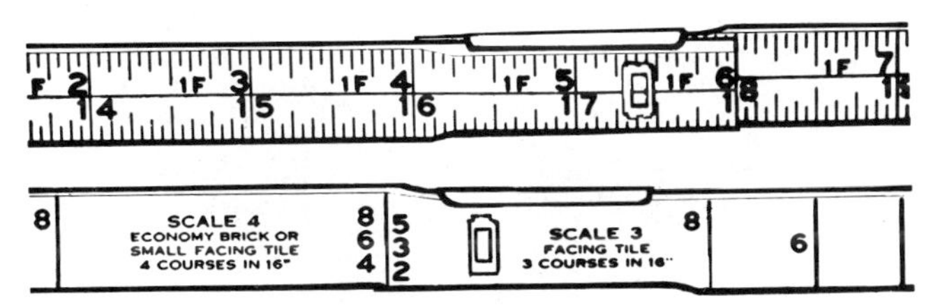

Inside scales on ruler (S)

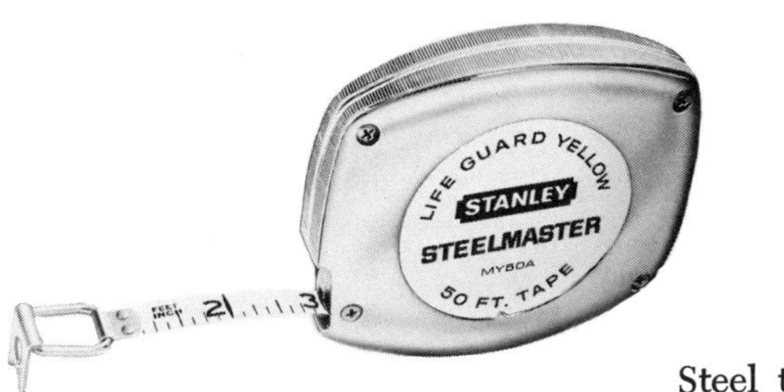

Steel tape (S)

Brick chisel (S)

Masonry wheel for power saw (S)

Appendix B

Appendix B shows many of the symbols used in blueprints. Each has a special meaning and purpose.

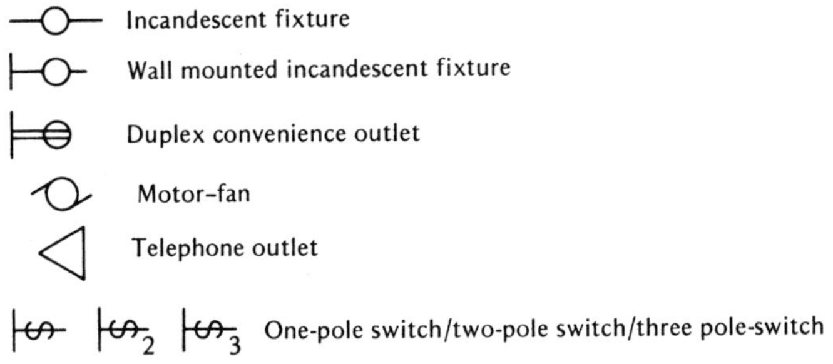

Incandescent fixture

Wall mounted incandescent fixture

Duplex convenience outlet

Motor–fan

Telephone outlet

2 3 One-pole switch/two-pole switch/three pole-switch

Electrical symbols

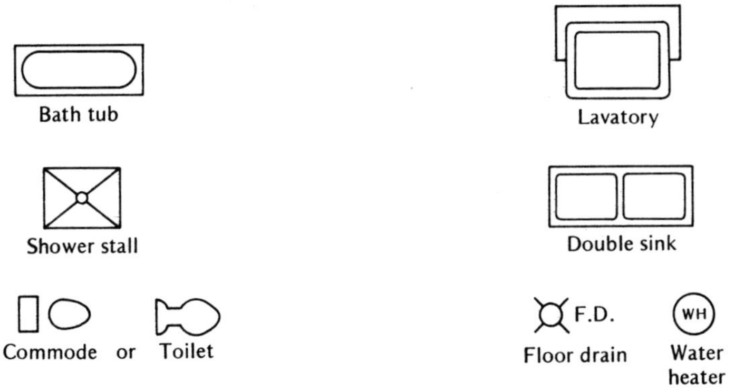

Bath tub

Lavatory

Shower stall

Double sink

Commode or Toilet

F.D.

Floor drain

WH

Water heater

Plumbing symbols

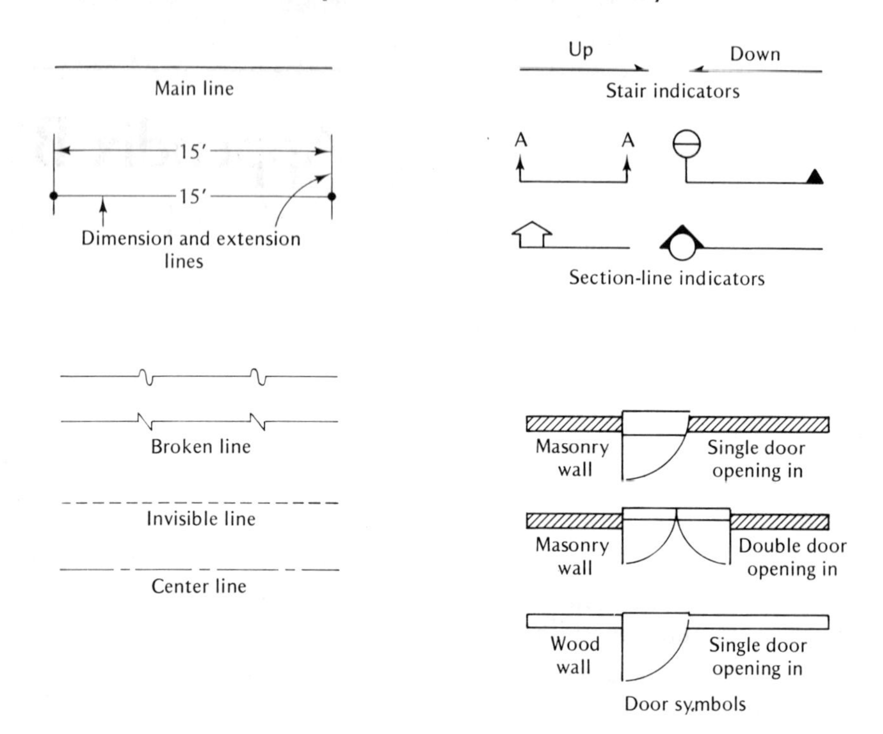
Main line
15′
15′
Dimension and extension lines
Broken line
Invisible line
Center line
Up
Down
Stair indicators
A
A
Section-line indicators
Masonry wall
Single door opening in
Masonry wall
Double door opening in
Wood wall
Single door opening in
Door sy,mbols

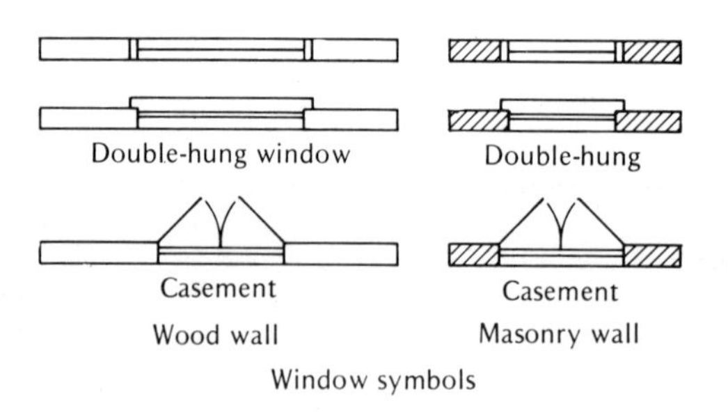
Double-hung window
Double-hung
Casement
Casement
Wood wall
Masonry wall
Window symbols

Foundation/floor plan symbols

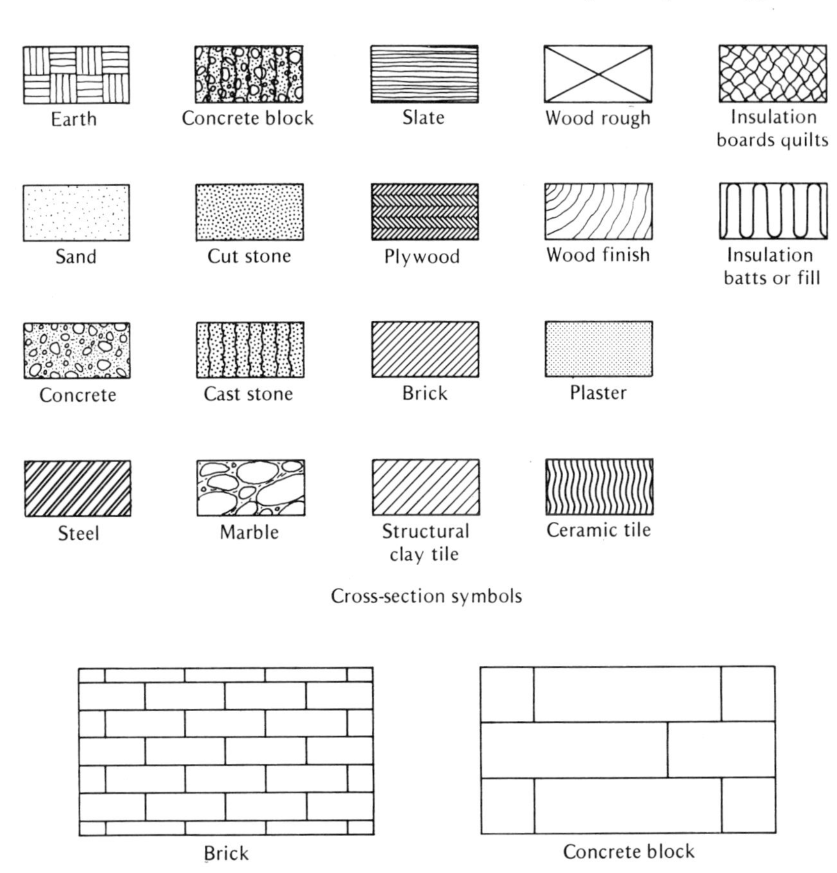

Cross-section symbols

Wood siding

Vertical wood siding

Elevation frontal symbols
(note: see other examples in text)

Index

A Good Mason Batch is
5 parts Screened Sand
1 1/2 parts Mason Cement
1/4 Part Porlant or Reg Cement

St ~~u~~cco 1st Coat - Spray/mist on wall
~~4~~ parts Screened Sand
1 Part Mason Cement
1 Part Porland or Reg Cement

Stucco 2nd Coat
~~4~~ Spray - 2 Parts Sand
1 1/2 Parts Reg Cement
1/3 Coffee Can Lime

Cement Batch
3 parts Sand
~~1~~2 Parts Rock
1 Part Cement